HANDBOOK FOR SURVIVAL

Information for Saving Lives During Radiation Releases and Other Disasters

ALLEN BRODSKY

Sc.D., CHP, CIH, Dipl. ABR

www.BrodskyBooks.com   www.Actions4Survival.com

PUBLISHED BY FIDELI PUBLISHING INC.

HANDBOOK FOR SURVIVAL
*Information for Saving Lives During
Radiation Releases and Other Disasters*

© Copyright September 2014, Allen Brodsky

All rights reserved.

No part of this publication may be reproduced, distributed, or transmitted in any form or by any means, including photocopying, recording, or other electronic or mechanical methods, without the prior written permission of the publisher, except in the case of brief quotations embodied in critical reviews and certain other noncommercial uses permitted by copyright law.

ISBN: 978-1-60414-810-7

For permission requests, write to the publisher at:

Fideli Publishing, Inc.
Attn: Permissions Coordinator
119 W. Morgan St. Martinsville, IN 46151

Ordering Information

Special discounts are available for quantity purchases by corporations, associations, and others. For details, contact the publisher at the address above.

Printed in the United States of America

Contact Information
Request for further information from Dr. Brodsky at
albrodsky@aol.com

or visit

www.Actions4Survival.com

Table of Contents

Acknowledgements vii
Preface ix

I. **Simple Actions to Take at the Moment of an Explosion or Storm to Save Life** 1
   Most Important to Know 1

II. **Understanding Dangerous vs. Safe or Zero Ranges of Radiation Dose and Risk** 5
   IIA. A Few New Words to Learn 5
   IIB. Estimating risks of exposures or doses under emergency conditions 14

III. **Ranges of Radiation and Concentrations of Radioactivity from Everyday Sources** 26

IV. **Further Preparations to Improve Chances of Saving Lives** 30
   IVA — Stocking Necessary Supplies to Last Until Normal Deliveries Can Be Expected 30
   IVB — Building or Planning to Use Shelters in Home or At Work 34
   IVC — Instruments or Dosimeters to Measure Doses or Radioactive Contamination 39
   IVD — Training for Immediate Aid to Blast and Trauma Victims 61
   IVE — Training and Supplies for Immediate Decontamination and Waste Removal 66

V. **How to Find Reliable Experts You Can Believe: Very Difficult But Doable** 69
   VA — The Bad News: The Difficulties and My Failures 69
   VB — The Good News: Opinions Detrimental to Health and the Environment Can Be Turned Around by Truth; How to Find the Information and the True Experts to Help 81
   VC — Factors that limit the risks from nuclear power plants with molten nuclear fuel 90

VI. Conclusions..93

APPENDIX A
Glossary..96

APPENDIX B
Methods for Checking Food and Water for Emergency Use...................113

APPENDIX C
Reasons for Urgency in Family Preparations to Save Lives...................121

AFTERWORD
The Good News About Nuclear Destruction by Shane Connor...............127

References...133

Index...139

About the Author..144

Flyer for 2011 Book..146

Table of Exhibits

EXHIBIT 1
Pocket 60-Second Training Card 2

EXHIBIT 2
Simple Immediate Actions to Reduce Injury and Radiation Dose 3

EXHIBIT 3
External Radiation Levels vs. Early Effects 17

EXHIBIT 4
Summary of Radiation Doses for Early Effects Ranging from
   Harmless to Lethal 18

EXHIBIT 5
Natural and Common Radiation Doses 27

EXHIBIT 6
Preparing a Makeshift Shelter with Planned Shielding Materials 34

EXHIBIT 7
Table of Approximate Densities, and Thicknesses for
   One-Fifth Gamma Reduction, of Common Materials 35

EXHIBIT 8
Protection Factors at Various Locations in a Variety of Buildings 38

EXHIBIT 9
An Affordable, Simple, Personal Radiation Monitor 46

EXHIBIT 10
Pocket Ionization Chamber and Static Electricity Charger....................48

EXHIBIT 11
Picture and Discussion of "High" Geiger Counter Readings and Chemical Color-Changing (SIRAD) Dosimeters vs. Possible Stay Times for Rescue or Seeking Shelter 55

EXHIBIT 12
An Affordable Geiger Counter with an Extremely Wide Range Satisfying Requirements of Every Response Organization and neighborhood 58

EXHIBIT 13
Kit of Radiation Instruments and Information, with Ionization Chamber Survey Meter in Upper Left Hand Corner 60

EXHIBIT 14
Ranges and Areas of Blast Effects from Nuclear Bombs 65

EXHIBIT 15
Limits on Food and Water Concentrations for Emergencies 114

Acknowledgments

I thank all who provided detailed comments on earlier drafts of this book. The comments were all helpful and resulted in important revisions. I will not mention the commenters' names, because they might not want to be identified with this final product. All facts and views are my own responsibility.

However, I would like to express deep appreciation to Robin Surface of Fidel Publishing for the attractive page-makeup of the book, and her patience with the many changes. She also made many important suggestions for content and editing, but she will also be protected from my identifying them in case her suggestions were not incorporated in the best way.

Preface

This book provides basic information to help the individual citizen take proper actions for survival, and avoid the most serious mistakes, in the immediate moments after the release of radioactive or toxic agents. A case is also made that suggested actions and information — such as immediate responses, appropriate sheltering and stocking of provisions — will also prepare for survival from natural disasters, as well as from radiological, chemical and biological agents.

Appendix C should be read carefully. It includes facts indicating the particular urgency for all citizens to understand and learn minimum actions in this book for saving the vast majority of lives in the event of attacks or events releasing radioactive materials. During the past year, you have all heard on the news about the increasing threats to our nation. Those who vow "Death to America" are getting closer to obtaining nuclear weapons, and already possess and can disperse chemical and biological weapons, despite all of our sanctions and negotiations. The ways that those who despise us can obtain nuclear or radioactive materials are described further in my book, by Allen Brodsky, *Actions for Survival: Saving Lives in the Immediate Hours after Release of Radioactive or Other Toxic Agents*, published by MJR Publications, Baltimore, MD, 2011. This book is referenced in the list of references at the end of this book as Brodsky (2011). A minimum of references is listed at the end of this book in alphabetic order by author, and by date, to provide readers who want to check author qualifications and facts supporting the recommendations in this book. I apologize for referencing my own books, but it is the easiest way for me to back up and document my facts, and also introduce interested readers to the vast literature on this subject.

Brodsky (2011) also has more detailed information useful for responders and scientists who arrive on the scene to assist in longer-term protection of the public, and the clean-up of areas for permanent residence.

> *"A prudent man foresees the difficulties ahead and prepares for them; the simpleton goes blindly on and suffers the consequences."*
> **— Proverbs 22:3**

I. Simple Actions to Take at the Moment of an Explosion to Save Life

MOST IMPORTANT TO KNOW

The following **Exhibits 1** and **2** provide a minimum of information that members of the public must know to protect themselves in the immediate moments and hours after an explosion to protect against nuclear blast effects and any radioactive that might be released. Some of these actions can be helpful whether a blast sound (explosion) is from bombs, accidents, or natural events such as tornadoes. These simple measures can save many lives, even without further knowledge about blast or other effects.

The first card in Exhibit 1, "60-SECOND NUCLEAR DETONATION TRAINING FOR FIRST RESPONDERS," is a small wallet card that has been distributed to thousands of responders in the USA, as well as recently (2013-2014) throughout Japan, by Steve Jones, representing Physicians for Civil Defense. He has volunteered for this life-saving task over many years of visits to fire and emergency department stations.

In Exhibit 2, I have revised the brief paragraphs in Exhibit 1 and extended them to apply also to "dirty bombs RDDs" that might be used by terrorists, and also to provide another way of estimating the reductions with time of fallout radiation from nuclear bombs. A card developed by the Homeland Security Committee of the Health Physics Society is also included in my 2011 book, but the simpler lists in Exhibits 1 and 2 are preferred here. The most important caution is that nobody should go to a window or near glass if an explosion is heard or bright flash of light is seen; the first thing to do is duck under something for cover and stay down for at least one minute. A blast wave travels after an instant at about the speed of sound, 5 seconds per mile. If you are 12 miles away, you will not be in the range of destruction of even a likely atomic bomb, but you might be cut by glass if you are near a window.

EXHIBIT 1

Pocket 60-Second Training Card

The bottom half of this card is on the front; the upper part is printed on the back of this card. The one in this picture may be reduced in size, cut out, and folded to imbed in plastic. See reference Jones (2014).

60-SECOND NUCLEAR DETONATION TRAINING FOR FIRST RESPONDERS

A.
Drop & <u>cover</u> when you see a flash. Stay down behind cover for two full minutes. Even covering with a newspaper can prevent burns. Keep eyes closed during bright light to prevent blindness.

B.
7/10 Rule: Fallout loses 90% of its radioactivity in the first 7 hours after a detonation and an additional 90% for every 7-fold increase in time: 90% in the first seven hours; 99% in 49 hours (two days) and 99.9% in two weeks.

C.
Fallout looks like sand, ash or grit as it falls and accumulates on the ground. If no fallout is visible on ground, there is no radiation! To be sure, place a piece of white paper, a dinner plate or anything with a smooth surface on the ground & check every 15 minutes for fallout particles

If visual indications of fallout appear take shelter for two or three days underground or behind thick walls. (These tips are generally true.)

This card may be the only nuclear training you get. Knowing A, B, & C, can protect your life and your department.

Give a card to each emergency responder in your jurisdiction.

These principles were developed during nuclear weapons tests from the 1940s-1960s and remain valid today!

The laws of physics do not change.

This card by: PhysiciansForCivilDefense.org

EXHIBIT 2
Simple Immediate Actions to Reduce Injury and Radiation Dose

(Some revisions and additions to Steve Jones' card as suggested by this author)

Post this on your refrigerator or reduce the bold words to place A to C on a wallet-sized card for regular review and ready access.

A. Drop and cover when you see a bright flash, or sound of a bomb. Do not go near a window that might soon shatter glass. Even covering with a newspaper will reduce burns from an atomic detonation, if outside the immediate areas of destruction. In the case of an atomic bomb, stay down for at least two minutes. Keep eyes closed during any sustained bright flash.

B. <u>7/10 rule for A-bomb radiation</u>: After all fallout deposits in your area on the ground, then every further 7 times the time it reduces at least to another tenth. That is, it loses 90% of its radioactivity in 7 times the time it deposited. If you are close-in enough and the intensity is 100 R per hour at 1 hour, then at 7 hours it will be 10 R per hour, and at 7x7 = 49 hours it will be only 1 R per hour.

Even at 10 R per hour, a family member could go out for one-half hour to obtain food and supplies and receive only 5 R — the permissible annual dose for a worker in peacetime. This would result in no acute illnesses and would at worst result in only a small fractional increase in cancer likelihood many years later.

<u>**For single or multiple radionuclides in an RDD, IND, or nuclear power accidents. Here, the 7/10 rule does not hold. Only the half-lives of the released isotopes that escape the protective building are important.**</u> (See likely isotopes in Brodsky (2011) and acceptable emergency intakes by inhalation, in addition to their half-lives, and further background on my estimates of acceptable concentrations in food and water for 10- and 30-day emergency consumption given in Exhibit 15.

C. Fallout might look like sand or grit falling down close in from a bomb, but might not be visual at farther distances. You may check to see if close-in fallout has arrived within minutes by observing accumulation of dust on smooth light surfaces, or by

a peak in the reading of a GM counter (see Glossary). But, do not go outdoors to do so!!

Better yet, get the easily read SIRAD dosimeters (stamp or card size in Exhibit 9 or 11), or the type of Geiger counter in Exhibit 11or 12, for your family from Shane Connor. Large purchases of SIRAD dosimeters should be directed to Dr. Gordhan Patel at sirad@jplabs.com, or JP Laboratories, Inc., 120 Wood Avenue, Middlesex, NJ 08846. Governments should be urged to distribute them.

II. Understanding Dangerous vs. Safe or Zero Ranges of Radiation Dose and Risk

"Happy is the man that findeth wisdom and gaineth understanding, for better merchandise is it than the merchandise of silver, or the gain thereof than fine gold."
— PROVERBS 6:6

IIA. FIRST, JUST A FEW NEW WORDS TO LEARN: FOR UNDERSTANDING PROTECTION MEASURES AND AVOIDING FEAR OF RADIATION UNDER EMERGENCY CONDITIONS.

YOU CAN SKIP THIS CHAPTER II AND GO ON TO CHAPTER IV TO PREPARE COMMON SENSE WAYS TO SAVE LIFE DURING AN EMERGENCY, BUT PLEASE: **TRY TO FIND TIME TO COME BACK LATER AND LEARN THIS INFORMATION TO PREVENT PANIC AND SAVE LIFE DURING RADIATION EMERGENCIES.**

Reasons for learning a few new words: Although some reviewers, including some of my scientific friends, have suggested that this book be written for a sixth-, or even first-grade level, I will not write for the lowest levels of intellect. I have experience in communicating radiation information to all levels of education — first-grade, eighth-grade, Ph.D., and all levels in between, including public-school and graduate-school classrooms, public forums, before lay juries, to police and firemen in atomic bomb fallout fields in Nevada, and more. Some of the things I have learned are presented in the early chapters of my 2011 book. I have found that those citizens likely to read and use the material in this book will have the motivation and ability to understand and use this information to protect themselves, their families, and their neighbors or co-workers. I do not want to dilute the facts and concepts that they will need in order to attract a few readers at the lowest levels of education and interest.

Understanding radiation doses and risks should not be as difficult as most people think. However, because reports from the news media often confuse or exaggerate radiation incidents, and our schools do not provide the basics to understand radiation issues, I must define a few quantities here to make the rest of the recommendations for protection understandable.

I will start by introducing each new definition in "plain language." Even for the very educated reader, it might take some concentration and review to learn the few new words needed to manage your protection from radiation, but certainly much fewer new words than you needed to learn in any courses in elementary school. Your new understanding will be worth it.

DEFINITIONS AND CONCEPTS USED IN DISCUSSING RADIATION ISSUES

Important Notice: I must start here with another explanation: why I am introducing two different units for each of the same quantities. The Board of Directors of the Health Physics Society (HPS) has instituted a ban on **"traditional units" (earlier SI units of R, rad, rem, etc.) of radiation quantities** in radiation protection practice, which have been in use for many decades. **I think that this ban is a big mistake.** The HPS has adopted the practice in its publications to allow the use only of **newer** (circa 1970-80) **Systeme Internationale (SI) units** recommended by theoretical physicists of the **International Commission on Radiological Units (ICRU)**, whom I believe could not have realized the disadvantages of removing the use of the traditional units that better relate the usual measurements of exposure in air in roentgen units to absorbed doses within the human body. Although as I document in one of the appendices of my 2011 book, members of the **International Commission on Radiological Protection (ICRP)** were originally uncertain about using the ICRU recommendations for unit changes and many did not favor it.

I have circulated a petition to reverse this HPS ban in the United States, for both scientific and practical reasons. I find that many of the scientists and practitioners who responded to my petition have indicated they will still use either or both sets of units. Some of our most outstanding scientists and physicians still have been using mixed units in some of their most recent presentations. **Also, some of the instruments that will be available in the event of an emergency will still use either or both sets of units.** Thus, I will use both sets of units and alternate them in some cases, so you will be familiar with them. Converting from one set to another can be done simply by using multiplying constants that will be presented, and that are easy to remember with just a little practice. **Just know that many who must be "politically correct" will condemn me and anything I write.** I just know that my open-minded readers will examine my evidence and experience as presented here, or upon further inquiry.

Radiation exposure — the amount of gamma photons (also called quanta or particles) or beta rays (free and fast electrons) **coming to you** per unit area of your body. Gamma radiation is the most penetrating and most likely to cause early or late effects at high quantities. Beta radiation coming from outside the body penetrates no more than about 1 centimeter beyond the skin surfaces. Levels of contamination on skin that will give high doses are given in my 2011 book. I know from experience working with soldiers in Nevada test fallout that skin contamination can easily be reduced to safe levels by brushing clothing and washing exposed skin with soap and water. Alpha rays (particles) from contamination on the skin cannot penetrate past the dead layer of skin from outside the body. In a very high range of **internal** body concentrations of alpha emitters, animal experiments have shown that cancers can be induced over the animal lifetime. However, autopsy data on deceased humans who were exposed in the United States to internal alpha emitters like plutonium in accident situations, have not shown any cancers from internal doses to the organs exposed, even though some of their doses to "critical organs" like bone have had cumulated dose calculated as up to thousands of rem, even before the latent period for induction of cancer. This is consistent with some animal experiments (see for example Raabe 2011). See also Chapter 4 of my 2011 book, supported by scientific literature references and my own experience. Alpha exposures are not an immediate concern here in emergencies with limited time exposures, despite the frequent reports that "plutonium is the most dangerous toxin of all".

Radiation instruments for monitoring the worker or public environment have sensitive probes that are used in "free air," and are not placed inside anyone's body. Radiation instruments therefore use the original roentgen units (R or mR), which can be related with certain physical factors to absorbed doses (see below) at various points in the body. When the entire body is immersed in a fairly uniform field of gamma or x radiation, the readings in air in R are often close to the maximum absorbed dose in body tissue in rad, and thus for ease in administration and conservative safety, the readings in R or R per hour are interpreted as whole body absorbed doses or dose rates.

You will not likely find radiation instruments using the newer SI units of exposure, which are coulombs per kilogram of air (C/kg). These new units could be dangerous if misinterpreted, because 1 C/kg deposited and measured in an air chamber instrument at STP would be equivalent to 3,881 R, a lethal exposure for only 1 unit (Cember and Johnson, pages 207-210). Therefore, field instruments using new SI units read in dose units of Gy or Sv, assuming measurements are taken inside the body.

Physicists and physicians working only in diagnostic or therapeutic radiology have now switched to using the new SI units, except that they use centigrays,(cGy) in place of rad. The cGy is a subunit of the gray that does not provide the theoretical benefit originally attractive to the ICRU theorists of having only a 1 in front of every equation. It does not matter to these radiology professionals unless they need to use instruments for field or laboratory monitoring. These days, if they are not involved also with surveys of laboratory safety with radioactive materials, they measure doses only within the body or human-like phantoms, or have other instruments for dosimetry calibrated to such in-body measurements, so they can be "politically correct" with today's ICRU and NCRP.

Radiation absorbed dose — the amount of **energy absorbed** per unit mass of tissue.

> **1 rad = 100 ergs per gram;**
> **or in the newer SI units, 1 gray (Gy) = 1 Joule per kilogram (J/kg),**
> and it turns out that
> **<u>1 Gy is equivalent to 100 rad</u>** (a good thing to memorize)
>
> **Thus, 1 rad is equivalent to 0.01 gray (Gy) in the newer SI units.** (another good thing to memorize)

In emergencies where the entire body might be exposed, the absorbed dose is assumed to be an approximate average over the whole body.

An historical point for the interested reader: For simplicity, dose in conventional units was very early rounded to 100 ergs per gram, because 1 R of exposure would deliver a dose of at most 96 ergs per gram to a **small volume of soft tissue in air** at the point of exposure. (A small volume is defined here so that no significant tissue around it would be assumed present to underestimate incident exposures; otherwise, appropriate calculations by scientists need to take into account the absorption in tissues shielding sensitive organs.). **Rather than go further here, just** know that stating exposure in R provides a high-sided estimate of the dose to tissues in the body in rad.

The same is true for biologically equivalent doses for radiations other than gamma or beta in **rem (which is 0.01 sievert (Sv))**. It is easy to convert between traditional SI units and the newer SI units by **multiplying or dividing by 100 (or moving decimal places)**, but care in checking such conversions is necessary to avoid dangerous mistakes. (**You might want to practice such conversions as you read through this book**.) Many current instruments, such as Geiger counters, can be switched to either set of units, just as a speedometer on your auto can be switched to read either miles per hour or kilometers per hour.)

Radiation biological dose — For specific organs or tissues within the body, "tissue weighting factors (w_t)," have been provided by the ICRU and ICRP to adjust for the same risk equivalent to a total body exposure when just the specific organs are irradiated. Multiplying the dose in **rad or Gy** by these factors gives values of equivalent (biological) dose in **rem or Sieverts (Sv), respectively**.

These weighting factors are useful for scientists who later take various measurements of radioactive material in the body, or in body fluids, to estimate internal doses to different body organs, which might have different chances of developing cancer or other effects, from the same doses in energy absorption from radionuclides breathed or ingested into the body. However, except in the rare cases when persons are caught in the radioactive (or "mushroom") cloud, internal doses are not likely to be as serious as external doses, and would likely be negligible for those indoors during times of maximum fallout and external radiation.

Therefore, for the general citizen's purposes in providing immediate protection, any measurements from radiation instruments should assume that:

1 R = 0.01 Gy = 0.01 Sv, and you will be on the safe side, or 1 Gy = 1 Sv = 100 rad = 100 R
(Again worth memorizing for immediate understanding of emergency situations.)

Just practice learning the above equivalents by moving decimals two positions. Obviously, using the newer SI units and thinking they are the traditional SI units could be very dangerous; another reason I advise learning both sets of units and being cautious with decimal points.

Radioactive material — any material (food, water, dirt, etc.) that contains any species of unstable atoms that emit ionizing radiation such as alpha, beta, or gamma rays. The unstable atoms are called "**radionuclides.**" Each of the hundreds of radionuclides we deal with has its own radioactive decay rate (transmutation rate) that is described by its **half-life** (see definition below). There are many naturally-occurring "radionuclides" that expose us every day, but these exposures are of low intensity, and it is believed that through the millennia, the human race has evolved with health improvements from these low-level natural exposures.

Half-life and radioactivity units — A particular species of radioactive atom has a half-life, which is the **time it takes for half of the atoms to change (sometimes called "disintegration" or "transformation")**. Half lives are given in units of time, seconds, minutes, hours, days, or years. Measured or calculated disintegration rates (sometimes called transformation rates), the rates at which atoms decay, are often given in curies (Ci), the newer SI units of bequerels (Bq) (which are disintegrations per second), disintegrations per minute (dpm), disintegrations per second (dps), or transformation rates per second (tps). Half-lives and disintegration rates of radionuclide concentrations in food or water have wide ranges over time, and wide rates in terms such as dpm per gram of pure nuclide, or Bq per kilogram of whatever materials with which they are mixed (Bq/kg or Bq kg^{-1}). Actual risks of given rates or doses depend very widely on which radionuclide(s) are in food, water, or already inside human tissues.

1 Ci = 37 billion atoms decaying per second; 1 Bq = 1 atom decaying per second. (Again, please memorize these new couple of words.)

(Each atomic nucleus of a specific amount of radioactive material has its own constant chance of decay over time, which leads to this phenomenon of half-life.) Inside the body, a radionuclide is like a tiny, submicroscopic x-ray machine emitting radiation (alpha, beta, and/or gamma) and bombarding nearby cells at close range, decreasing in intensity at its own rate. However, there is also a physiological elimination rate, which is sometimes simply described as a "**biological half-life**". The biological half-life, or sometimes a changing rate of biological elimination from specific chemical forms absorbed into the body, is also taken into account in assessing long-term doses and risks of radioactive materials taken into the body by inhalation or ingestion. (See NCRP and ICRP reports, and Federal regulations, Title 10 Code of Federal Regulations, Part 20 (10 CFR 20), on these calculations and limits of intake. Some are referenced in Brodsky (2011 or 1996).

NOTE: Nuclear bombs release the entire spectrum of radioactive fission products produced at the time of explosion. On the other hand, nuclear power reactor accidents release only a few of the isotopes of more volatile elements such as Iodine-131 and Cesium-137, because there are many filtering or deposition stages in pathways for radioactive materials to reach the environment in USA-built and -regulated reactors. I-131 concentrates in the thyroid, which requires at least 20 times the dose in rad (or gray) to produce the same chance of cancer as the same dose to the whole body (if interested, see (Brodsky 2011, page 753)). Cs-137 concentrates in whole body muscles so its dose rate is diluted.

Many, like our son, have received medical radiation doses to their thyroids of thousands of rad (thousands of cGy, which are millions of millirad (mrad)), and they are cured of cancer. Their long-term chances of later cancers from their therapeutic treatments are negligible compared to life-saving benefits of treatment. Regarding doses from Cs-137 intakes from reactor accidents in the USA, none were even measurable after the Three-Mile-Island accident, covered elsewhere in this book. All this is true, despite the many scary stories in the media. In the United States, accidents or mishaps in U.S. nuclear power plants have released extremely small fractions of the radioactive materials present in the reactors.

(See some reasons for relatively low emissions in sub-section VC later in this book.)

Still, in (Brodsky 2011), some guidance is given as if all radioactive material is released in an attack or accident. Then, the scientist or person evaluating the incident can apply the actual fractions of material determined to be released, in order to estimate doses or risks to the public. (However, see the misinformation in the news article after the Three Mile Island accident, pictured in Exhibit 3 of my 2011 book, and my surrounding evidence in the 2011 book about how Dr. Sternglass circulated an effective million-sized lie in AP press articles all over the nation by not considering fractions released, panicking much of the public.)

Prefixes — Because such wide ranges of quantities are discussed, prefixes often precede units, such as: µBq means millionths of becquerels; mBq means thousandths of Bq; kBq means thousands of Bq; MBq means millions of Bq; and GBq means billions of Bq. The same prefixes are used in stating quantities of exposure or dose. Study them if they are not familiar. Below is a table of prefixes in case you run across them in the media.

SI PREFIXES

exo (E)	— 10^{18} (a billion billion)		deci (d)	— 10^{-1} (a tenth)
peta (P)	— 10^{15} (a million billion)		centi (c)	— 10^{-2} (a hundredth)
tera (T)	— 10^{12} (a thousand billion)		milli (m)	— 10^{-3} (a thousandth)
giga (G)	— 10^{9} (a billion)		micro (µ)	— 10^{-6} (a millionth)
mega (M)	— 10^{6} (a million)		nano (n)	— 10^{-9} (a billionth)
kilo (k)	— 10^{3} (a thousand)		pica (p)	— 10^{-12} (a trillionth)

or example, a gigaBq = a billion Bq, i.e., a billion dps;
a cGy = a hundredth of a Gy = 1 rad, just as a centimeter = a hundredth of a meter.
a µCi = a millionth of a curie (Ci), which is 37 billion/1,000,000 = 37,000 dps (37,000 Bq)

The above terms will be helpful in estimating your risks (or relative safety) of various levels of radiation exposure as presented in the following section, and in understanding announcements about large, small, or insignificant radiation risks of exposures or doses stated in the media or by local responders or authorities.

IIB. ESTIMATING RISKS OF EXPOSURES OR DOSES UNDER EMERGENCY CONDITIONS

IIB-1. Early Acute Effects: There are **early (or "acute") effects** of relatively high doses, shown in detail in ranges of whole body doses for adults in **Exhibit 3**. This table was constructed by physicians based on experience with the limited number of accident cases in nuclear processing plants, and from the Chernobyl accident, and is copied from my 2011 book. (Despite some media reports, nobody was killed, either workers or members of the public, from any nuclear power plant accidents in the USA, including that at Three Mile Island.) A **brief summary**, easier to remember, of the approximate dose ranges of these early effects is presented in the following **Exhibit 4**. Children and pregnant women should be limited to doses well below 5 rem if at all possible. Effects on the unborn are reviewed in my 2011 book.

One accident in the Gulf Research facility in the 1960s, in which I measured doses on a mock-up phantom placed in the exposure position of a worker to radiation from a Van de Graaff high-energy accelerator, was reported in (Brodsky and Wald 2004). The measured average dose to bone marrow obtained within two days after the accident was estimated as 600 rad (or rem, or 6 Gy), then confirmed by estimates from rings and dicentrics in the chromosomes of white blood cells by Dr. Niel Wald. Dr. Wald recommended a bone marrow transplant from an identical twin brother at exactly the right time, which saved the life of the exposed worker. Our experience with this patient can be seen to be consistent with the information in Exhibit 2.

The 600 rad dose above was an average dose to bone marrow all over the body. My measurements of doses to the hands and forearms were about 6,000 rad each. My measurements of the doses to the lower legs were 3,000 rad. Although the bone marrow transplant from the identical twin brother saved the worker's life, he then needed quadruple amputation.

This is an interesting case, managed by my physician supervisor Dr. Niel Wald, in the latter 1960s. It is worth noting not only because of the interesting aspects of such life-saving treatments, but also because many millions of dollars are still being devoted by our homeland security agency for medical researchers to improve knowledge of such life-saving procedures. I am not against research, but must advise that in the situations likely to exist after an accident or attack serious enough to expose many persons in populated areas, it is unlikely that the scarce physicians knowing such procedures would be able to find and treat the more highly exposed persons and save their lives. Therefore, there needs to be available on persons in the public the types of inexpensive SIRAD dosimeters, presented further in this book and in detail in Chapter 8, pages 198-206 of Brodsky (2011), that will warn persons if they are accumulating such high doses to seek better shelter, or if they have already received such high doses to seek medical treatment if available. Such dosimeters would also show the vast majority of the public involved with some levels of radiation that they are not in the range where serious health effects would occur, so they would not panic and rush outside for an improper attempt to evacuate.

In 2005, I met with Dr. Charles E. McQueary, a.high official in the Department of Homeland Security (DHS), who at the time seemed to appreciate the need to distribute inexpensive dosimeters to the public. He ordered the SIRAD dosimeters tested for accuracy and functionality at the DHS New York research laboratory. Despite the success of the tests (Buddemeir 2007) and the approval of DHS for dosimeter distribution, further officials of DHS failed to advise the States to order them at the relatively small costs of millions, while at the same time spending billions of dollars on fire engines and other familiar responder equipment, which would likely be of little use in saving lives of a public without needed information, panicked and blocking inappropriate evacuation routes.

Thus, it seems that it might take demands from members of the public, like you the reader, to obtain the personal dosimeters of proper range to empower yourself and local responders to act protectively in moments following nuclear or radiological events.

Information for estimating the **chances of late ("chronic") effects, such as cancers that might appear years after exposure to radiation, is presented in the next section, <u>after</u> the following Exhibits 3 and 4.**

EXHIBIT 3

External Radiation Levels vs. Early Effects

Early effects: Effects from short-term doses to the entire body, doses that are not compensated by the repair that can take place when exposures are drawn out. 1 Gy = 100 rad or 100 rem.

Symptoms Listed Below	Degree of severity and corresponding dose range				
	Mild (1-2 Gy)*	Moderate (2-4 Gy)	Severe (4-6 Gy)	Very severe (6-8 Gy)	Always lethal (8 Gy +)
Vomiting onset	2 hours after dose	1 to 2 hours after dose	Earlier than 1 hour after dose	Earlier than 30 min. after dose	Earlier than 10 min. after dose
% Incidence	10-50%	70 to 90%	100%	100%	100%
Diarrhea Onset	None ———	None ———	Mild 3 to 8 hours	Heavy 1 to 3 hours	Heavy Within 1 hour
% Incidence	———	———	Less than 10%	More than 10%	Almost 100%
Headache Onset	Slight ———	Mild ———	Moderate 4 to 24 hours	Severe 3 to 4 hours	Severe 1 to 2 hours
% Incidence	———	———	50 %	80 %	80 to 90%
Consciousness Onset	Unaffected	Unaffected	Unaffected ———	May be altered ———	Unconsciousness for secs. or mins. Seconds/mins.
% Incidence	———	———	———	———	100% @> 5 Gy
Lethality	0%	0 to 50% Onset 6 to 8 weeks	20 to 70% Onset in 4 to 8 weeks	50 to 100% Onset in 1 to 2 weeks	100% Lethal in 1 to 2 weeks
Medical Care Needed	Outpatient observation	Observe in general, treat in special hospital, if needed	Treat in special hospital	Treat in special hospital	Palliative treatment only**

* 1 Gy = 100 rad

** With appropriate supportive therapy, individuals might survive for 6 to 12 months with whole body doses as high as 12 Gy. The worker with an effective dose of 600 rem in Brodsky and Wald (2004), after a bone marrow transplant and quadruple amputation of forearms and lower legs that were in the beam,, died at age 55 from heart failure, not from radiation-caused cancer. His identical twin unexposed brother also died of heart failure at age 55.

This table was adapted from Tables 1 and 3 in F. Mettler, A. K. Gus'Cova, and I. Gusev, "Health effects of those with acute radiation sickness from the Chernobyl accident," Health Physics, Vol. 93(5):462-469; 2007.

EXHIBIT 4

Summary of Radiation Doses for Early Effects Ranging from Harmless to Lethal

Simple table of external radiation levels vs. acute effects, for "whole body exposure":

ACUTE RADIATION EXPOSURE

0-25 rads:	No observable effect
25-100 rads	Slight blood changes
100-200 rads	Significant reduction in blood platelets and white blood cells (temporary)
200-500 rads:	Severe blood damage, nausea, hair loss, hemorrhage, death in many cases
>600 rads:	Death in less than two months for over 80%

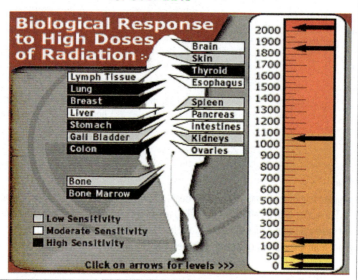

Note: 200 rad = 200 rem = 2 Gy = 2 Sv for our purposes.

IIB-2. Long-Term Effects: For the majority of those likely to be involved with measurable amounts of radiation, **various cancers** might occur years after exposures to amounts of radiation yielding doses from external whole body irradiation above about 20 rem (0.2 Sv). Estimating the risks of cancer will suffice for this handbook, because other late radiation effects, including hereditary effects, have been found to be small relative to cancers, and the lifetime detriment of these other effects has been absorbed by ICRP and NCRP recommendations into the cancer risk factors for our practical purposes.

Estimates of the chances of radiation-caused cancers as a function of dose were made by scientists and physicians examining over the years medical data and dose estimates of those exposed to the atomic bombs over Hiroshima and Nagasaki in 1945. They also used animal experiments in making risk estimates. My view of the best estimates of cancer risks (chances of getting cancer) for higher doses are those published in ICRP reports from 1977 to 1990 by the work of Dr. Warren Sinclair, who served as Chair of NCRP for many years following the retirement of its founder Lauriston Taylor. The many demographic translations and mathematical models used by Dr. Sinclair to obtain risks for U.S. citizens from the Japanese survivors are too complex to present here. (These methods may be examined by those interested in this science as reviewed in Brodsky (1996), or in the original ICRP reports referenced there.) Only summary estimates of risk are given here, which use high-sided estimates of the effectiveness of long-term exposure compared to that of the prompt gamma and neutron exposures encountered at Hiroshima and Nagasaki.

These estimates are also made using a **linear dose-response model without (no) threshold (LNT model)**, for purposes of deriving peacetime dose limits for managing radiation exposures to workers and the public under controllable conditions.

Moreover, these ICRP and NCRP estimates do not take into account the evidence for **hormetic effects** at lower dose ranges (protective, adaptive, and beneficial effects observed at acute dose levels below about 20 rem, and at low and much higher levels spread over long time intervals).

Dr. T. Donald Luckey in 1980 and 1991 published comprehensive reviews of thousands of references to published scientific articles in which lower level radiation exposures were found to be beneficial to the life of organisms consisting of only single cells up to animals and humans (Luckey 1991). He coined the term **"hormesis,"** which is derived from the same Latin word as hormone, because Dr. Luckey found that lower doses stimulated beneficial biochemical and physiological reactions in many species. Dr. Luckey was Chairman of Biochemistry at the University of Missouri. (Bless his soul, Don Luckey just passed away at 94 in March 2014.)

Those interested in the evidence of hormetic effects are referred to only a few books and articles here, which would lead them into further literature (Cuttler 2013, 2014; Luckey 1991; Goldberg 2009; Hiserodt 2005; Rockwell 2003; Brodsky 1996; and my update of Chapter 4 of the 2001 book included in Chapter 4 of Brodsky 2011). Books other than mine also put radiation risks in better perspective with other risks.

However, although no further information is given here about hormetic effects of low or extended doses of radiation, consider, for the time being, how selenium is now found to be a necessary element for health in small concentrations (see the information on your bottle of vitamin pills). When I was on Enewetok as an Army physicist working on the first hydrogen bomb test in November 1952 at age 24, I consulted an Army physician for something to remove my dandruff; I was already worried about losing my hair. He gave me a prescription for Selsun and warned me not to get any in my eyes or inside my body. The smallest quantities were deemed poisonous. Now, in 2014 and for many years, I have been able to buy hair products with selenium over the counter without a prescription. Selenium (Se) is now in my vitamin pills and considered an essential nutrient in the amounts in the pills. I thus have still not been poisoned, but am now benefiting for many years from the hormesis of the lower amounts of Se.

References in the paragraph above will give many other examples of beneficial hormetic effects at lower doses, of substances that are poisonous or harmful at higher levels of intake.

Maximum risk factor from LNT model: Note: The word risk is not used here always to imply danger, but only as another word to mean chance (of cancer, etc.) See Glossary. The word risk may be used here, for instance, to say, "The risk of any health effects is zero; there is no danger and the risk might be negative and mean that there are hormetic or beneficial effects at these dose levels".

The following estimate of the long-term chances of cancer deaths shows the relatively low proportional increase in death from cancer in a responder (or citizen saving a family) if under the recommended limit for emergency exposure of 25 rem (0.25 Sv) acute dose, assuming hormetic effects are not considered to be effective at this dose level. The estimate is probably high-sided, but is taken from information in a book by two of my very competent friends (Cember and Johnson 2009, page 327), who reference, for use above about 10 rem, the risk factors used by the International Commission on Radiological Protection (ICRP 1991). The following sentences in italics are my own adaptation:

"*The **maximum** estimated chance of dying from long-term cancer for chronic gamma exposures is 5 chances per 10,000 rem (5 chances in 100 per Gy). This means that if you receive a general whole body dose of 25 rem during a rescue or to get food, during the first day of an incident, you will have at most a chance of 25x 5/10,000 = 0.0125 chance of dying of cancer later in life. This chance is the same as 1.25 percentage (%) points. The current rate in peacetime from all causes in the USA is 24%.*

Therefore, using this risk factor, a 25 rem dose would increase your natural risk of 24% to 25.25%, but these calculations can not be assumed accurate for a particular person. (See definitions of "risk" and "stochastic effects" in Glossary.)

Remember, as cited in the references discussing hormetic effects, as dose is lowered down toward 10 rem and lower, protective effects come in at different levels for different individuals, and these protective effects can not only lower the chances of later cancers, but also promote immune system activity that even decreases the chances of other serious diseases.

Nevertheless, the above risk factor used at 25 rem can be multiplied at that level and at higher acute doses to estimate chances of late cancers when higher doses must be received for the saving of life. Certainly, most of us at an adult age would readily take such a risk of late cancer in order to save a member of family, or even a neighbor, since this dose is below where any immediate ill effects would occur or be felt (**Exhibits 3 and 4**). Note, moreover, that when older persons are available for rescue or life-saving, the **oldest able person should volunteer to perform the rescue, since the latent period for cancer to develop may reduce the chances of cancer during his lifetime.**

One further reminder: We all know of friends or members of our family who have undergone radiation treatments to enhance the possibilities of removing cancers. For breast cancer treatments, for example, a total dose of about 6,000 rad is given in 200 rad per day treatments, five days per week, for about six weeks. Each treatment, for someone in my family, is delivered in a moving beam at 600 rad per min, for about 20 seconds, from a linear accelerator. This spreading out of this large 6,000 rad dose spares injury to healthy tissues more than to any still-encapsulated tumors, so these radiation treatments are often successful in curing the cancer.

Now, referring back to the previous section on acute effects, we see that the dose to the hands and forearms happened to be measured as 6,000 rad, and the dose to the lower legs as 3,000 rad, for the worker who then needed quadruple amputation (Brodsky and Wald 2004). The exposure time of the worker, estimated from the radiation beam intensity, was also about 1.5 minutes. Thus, it is clear that the spreading-out of doses even as high as 6,000 rad over six weeks, although given in 30 acute 30-second doses of 200 rad each for six weeks, prevents complete destruction of breast, lung, and heart tissues that are in the beam or its penumbras. Such a 6,000 rad dose would lead to death of the patient if the 6,000 rad was given in one treatment time rather than 30 over six weeks.

As recent literature on radiation effects shows, this lowered risk of doses spread over time also applies to the long-term cancer effects of even the higher doses that might be received after an incident. However, the "expert" bodies that recommend limits for peacetime exposure of workers and the public use only one conservative factor of two to reduce acute risk to the risk factor of 5 chanegs per 10,000 rem given above for use in long-term exposures, and still assume that the same resulting risk factor applies no matter the amount of time over which the exposure is received. These expert bodies, also perhaps for simplicity in managing peacetime radiation protection programs, also assume that the effects of all doses are cumulative, no matter the amount of time over which they are spread. These doses would not be cumulative when received over long times when the tissues are subject to repair, or when hormetic effects occur at the lower dose levels.

Regulatory agencies in the USA and elsewhere have constrained themselves over the years to follow recommendations of the ICRP and NCRP in writing formal regulations consistent with the recommendations of these expert bodies. Thus, it can be expected that some authorities in the USA will not be familiar with the literature quoted in this handbook, and might object to my recommendations. But note here, they will be wrong. When radioactive material is released in large amounts in an **uncontrolled** manner, the best scientific knowledge of radiation effects, including hormetic effects, must be used to prevent panic and save lives.

Bottom Line: For acute doses above 20 rem for an individual adult of average age, the LNT-estimated risk factor of 5 chances per 10,000 rem (cSv) may be used to obtain an **upper limit** of risk of a later cancer.

Note however, that a scientist whom I know to be outstanding and reliable has recently examined much of the relevant literature over the past 100 years (Cuttler 2013a,b,2014a,b). He suggests, based on his review, that about 70 rem/year (at 0.2 rem per day) is a level above which some elevation of cancer risk can occur, but below which hormetic effects are likely to overcome them. I have checked some of his references and he also shows that his conclusions are similar to those made earlier by Dr. Lauriston Taylor, founder of the NCRP, who had originally supported this dose and dose rate as a reasonable NCRP standard that would produce no ill effects.

However, remember that a given dose, when accumulated over a longer time, presents a **smaller** risk the **longer** the time over which it is accumulated, either for relatively acute or long-term effects. When these doses total below about 10 rem, there will also likely be protective effects that not only cancel the risks calculated by the above risk factor, but also stimulate hormetic effects that protect against other diseases.

Raabe (2011) has shown, with extensive collections and analyses of his own data as well as those of others, that when internal doses are spread at low dose rates over a lifetime, there are effective thresholds (which he calls "virtual thresholds"), so that total internal doses of up to 1,000 or more rem do not result in late cancers, even for radioactive species that are considered to be those that are the more dangerous — such as plutonium and strontium-90.

WHY WE SHOULD *NOT* BELIEVE THAT "ANY RADIATION IS DANGEROUS"

We have already discussed the evidence that shows, even without hormetic effects, at low doses or dose rates that might accumulate what have been considered high total doses by some current radiation standards, Raabe's experiments show that there is an effective threshold for inducing cancer during the lifetimes of animals (with other evidence (e.g., Cuttler 2013,2014) showing even beneficial effects at such dose levels). However, there are still some official government documents that state, "There is no safe level of radiation dose," which is the same as saying, "Any radiation is dangerous." As we have shown in the italicized paragraph above, even with the LNT hypothesis and without recognizing the evidence for hormetic effects, an acute radiation dose, or even a long-term cumulative dose, as high as 25 rem, or even higher, does not increase the prevalence rate of lethal cancers more than a fraction of a percent. (Using a dose-and- dose-rate-effectiveness factor (DDREF) of the ICRP has only reduced the estimate of long-term cancer risk by a factor of two.) Therefore, even a 25 rem dose should not be called "dangerous"; this would scare even responders who would otherwise feel justified receiving such a dose to save lives. I am afraid that much of our official responder training will scare responders unduly.

The next section presents some doses received from various activities that are familiar to anyone. It might be useful to examine the risks of each activity or medical treatment, using the risk information above in this section. In the case of a medical treatment or exposure to only a portion of the body, a rough estimate of the risk may be obtained using the estimated fraction of the body exposed, as well as the general high-sided risk factor. More specific weighting factors are provided by the ICRP and NCRP, but this detail is not necessary here for most citizens.

III. Ranges of Radiation and Concentrations of Radioactivity from Everyday Sources

Radiation doses from natural sources: Knowing the meanings of the terms above, you can now just examine the following **Exhibit 5** in this book to be familiar with the doses of natural levels of radiation and radioactivity that are present around you in ordinary everyday life. If you care to check references, you can also check and compare radiation and radioactive contamination levels from nuclear accidents, or from likely or possible terrorist attacks. (Buddemeir et al. 2011; Brodsky 1982, 2011)

Radioactivity levels from natural sources: A discussion of radioactivity in natural sources is presented after the **Exhibit 5** on the following page, for comparison with those that might be measured after an accident or terrorist attack

Ranges of Radiation and Concentrations of Radioactivity from Everyday Sources

EXHIBIT 5
Natural and Common Radiation Doses

TYPICAL RADIATION DOSES
Annual dose from living near a nuclear power plant <1 millirem

Flight from Los Angeles to London	5 mrem
Annual public dose limit	100 mrem
Fetal dose limit	300 mrem
Barium enema (limited part of body)	500 mrem
Annual radiation worker dose limit	5,000 mrem
Heart catheterization (skin dose)	45,000 mrem
Life-saving actions guidance (NCRP-116)	50,000 mrem
Mild acute radiation syndrome (if whole body dose)	200,000 mrem
LD 50/60 for human (bone marrow dose) (whole body dose)	350,000 mrem
Radiation therapy (localized & fractionated)	6,000,000 mrem

(Here, all doses except the medical diagnostic doses of barium enema, heart catheterization, and the cancer therapy doses, are assumed to be delivered over the entire body, which is the usual assumption when discussing fallout doses from nuclear accidents or bombs. **Remember that 5 mrem = 0.05 mSv; 200 mrem = 20 Sv; and 6,000,000 mrem is the same as the 6,000 rem (or rad for our purposes) delivered only to breast in cancer treatment, as discussed in Section II.)**

These numbers represent some USA natural average background and commonly encountered doses, and some regulatory peacetime "limits" (which I have helped promulgate, but which do not indicate thresholds of harm). Similar dose ranges would be applicable in Japan. These numbers can vary somewhat depending on your own activities and where you live. Population medical exposures have been widely reported to have increased due to CT scanning, but unfortunately these reports, and exaggerated estimates of harmful effects, have not been presented in context with lives saved. This is unfortunate, because many citizens have avoided CT scans and other radiology exams that could have saved their lives. The doses to the public in the vicinity of a nuclear power plant in normal peacetime operation are so low as not to be measurable; only estimates are given at the top of the table. In contrast, some dose rates inside the failed nuclear plants in the Three Mile Island, Chernobyl, and the Japanese disasters, have been so high that workers close to the reactors have needed to take turns limited to minutes in order to carry out possible corrective or recovery tasks. This shows the need to understand the effects of radiation doses and intensities over a very wide range.

Radioactivity levels from natural sources: About 70 of 340 nuclides found in nature are radioactive. Potassium-40 (K-40) atoms are a small fraction of the potassium atoms that enter the body with salt intake, and are needed in our bodies to maintain our heartbeats. The K-40 cannot, and need not, be separated from the non-radioactive K-39 atoms. The dose delivered is not a risk (such a dose is NOT dangerous). The internal dose from K-40 is the highest internal dose from naturally occurring radioactive material taken into the body, except possibly for the theoretical calculations of dose to lung tissue from radon daughters. Although these K-40 atoms have a half-life of 1.26 billion years, and each atom upon decay emits 1.46 MeV gamma rays only 11% of the time, there are so many of these atoms in our bodies that the internal dose per year from the beta and gamma radiation is in the range of 10 to 20 mrem per year, compared to our overall effective dose from natural background and radon daughters of about 300 mrem per year. (I have often had my students do a homework problem to calculate their doses from natural K-40.)

The gamma radiation from the K-40 is so easy to detect with modern radiation detectors that, when I was Technical Director of Radiation Medicine at Presbyterian Hospital (under Chairman Niel Wald, M.D., of the Radiation Health Department, Graduate School of Public Health, University of Pittsburgh) I used my own body sitting in a lawn chair inside our counting room (called a "whole body counter") at the hospital to calibrate my own equipment. Details of this experience are in the references (Brodsky 2011).

By comparison with the dose from natural potassium, I was also able to measure the amounts of cesium-137 (Cs-137) in my body from worldwide fallout from hydrogen bomb tests in my whole body counter. Agreements with the Soviets ended the atmospheric testing of nuclear weapons in 1963. Thus, peak amounts of fallout Cs-137 in our bodies occurred in the USA population in the 1963-64 era; about the same concentrations must have occurred in the Japanese population about that time.

Our whole body counter at the University of Pittsburgh opened for operation in 1964-1965, for checking persons in the Pittsburgh area who had inhaled plutonium, americium, and fission products from the then-blooming nuclear-related operations in Western Pennsylvania (see Brodsky and Wald (2004). Soon after our operations began, I measured about 3 nanocuries (nCi) total of Cs-137 in my body, mainly distributed in my muscle tissues. I do not remember the dose-rate to my body exactly, but it was certainly well below several millirem per year in 1964. The radioactivity in my body decreased each year thereafter due to the 30-year half-life of Cs-137 and the relatively short physiological half-life of cesium, as well as the decreasing amounts entering the body from fallout after cessation of the atmospheric nuclear tests. A few millirem per year is nothing to fear.

Certainly, the emissions from the Japanese reactors, without the stratospheric transport and much lower radioactive content than the hundreds of detonated nuclear bombs, could not affect anyone in the United States, and authoritative reports so far indicate that neither will the Japanese be seriously affected by radiation from the reactor emissions at Fukushima, compared to the extreme suffering caused by the tsunami. See Cuttler (2013).

In addition to cesium, the iodine nuclides are among the most volatile; I-131 and other iodines that might be inhaled if within hours of release, contribute doses to the thyroid gland. After the Windscale accident in England, which released about 20,000 curies (Ci) of Iodine-131, a limit of 0.1 microcuries per liter of milk was set to limit infant thyroid doses to less than 20 rem (Brodsky 2011; 1960).

(The 2011 book shows that this British limit happens to be consistent with the safety factor of 20 that I recommended in general for infant exposure compared to adult limits, in my review of Chapter 4 in the 2011 book. The Washington Post article by Andrew Higgins (2011) indicated that a farmer near the Fukushima site who refused to evacuate and continued to consume milk and food from his farm showed no internal radioactive material at all within his body when measured, even from the early effluents from the damaged nuclear reactors.)

IV. Further Preparations to Improve Chances of Saving Lives

IVA — STOCKING NECESSARY SUPPLIES TO LAST UNTIL NORMAL DELIVERIES CAN BE EXPECTED

The serious survivalist who is willing to spend time and effort for maximum protection has extensive information available for optimum protection of himself and family. This information, in many studies over many years, has been developed by civil defense and other survivalist organizations that can be found on the internet by searching the words "survival" or "survivalist." The Department of Homeland Security (DHS) on ready.gov has many recommendations for stocking food and emergency supplies; the DHS website also provides a checklist for a Family Emergency Plan in case family members are separated, and provides questions and answers that convey appropriate preparations for dealing with a radiation emergency. Much of this information is assembled starting on page 152 of Brodsky (2011), which can be downloaded inexpensively as an e-book from Amazon.com.

Many more articles with details and explanations for further preparations may be obtained from the 1968 to 2014 issues of the Journal of Civil Defense, available to members of The American Civil Defense Association; membership is available at a very low cost at the website www.tacda.org. The Journal of American Physicians and Surgeons (J Am Phys Surg), from which the trauma care information from Hatfill and Orient (2013) is summarized in a further section of this handbook, also has many articles helpful for comprehensive preparations to survive either natural or man-made disasters. The J Am Phys Surg can be obtained from the Association of American Physicians and Surgeons, Inc. (AAPS), by contacting www.jpands.org or (800) 635-1196

Regardless of the time and effort available for preparations, every family and citizen should be able to prepare the most essential items needed for an extended emergency by noting the items and medications used every day for a couple of weeks. Then, they should purchase these items from local stores BEFORE an emergency (hurricane, accident, etc.) occurs, because stores will run out of items rapidly if a disaster is expected to be imminent. The following is a short list of the most essential items:

- Water, one gallon per day per person for at least three days, and preferably at least two weeks, for drinking and sanitation. This can be obtained by purchasing a limited number of bottles of water, and also by having a clean tub available to fill before cutoff of supplies, and learning how to obtain water from your water heater. Inexpensive filter equipment available in local stores should be purchased, if not already owned, for all drinking water.
- Food, non-perishable, such as canned or bottled vegetables, fruit, chicken-vegetable soups. Most families should probably plan to use only foods that have been sheltered within their homes, or are in closed cans or bottles that can be washed with soap and water before being opened.
- Battery-powered or hand-cranked radio and a NOAA Weather Radio with tone alert and extra batteries for both.
- Flashlights and extra batteries. Light is needed to allow reading and prevent depression. Avoid candles and open fire; they use up oxygen needed for breathing in a tight room.

- First Aid kit. Also, see the need to train for immediate care of trauma victims in section IVD.
- Whistle to signal for help when needed.
- Dust mask, to filter contaminated air, and plastic sheeting and duct tape to seal shelters (although as I explain in this book and my 2011 book, the risks of breathing hazardous amounts of radioactive material are extremely low, and ordinary dust masks will not likely protect against toxic chemicals or biologicals. Still, being indoors in a closed room will likely provide considerable protection in either case.
- Moist towelettes, garbage bags, and plastic ties for personal sanitation.
- Chemical toilet, or improvised toilet made with 6-gallon can, toilet seat, garbage bags, perfumed detergent, draped and placed near entrance of shelter.
- Wrench and pliers to turn off utilities, or obtain water from heater. Crow bar in case anyone close to a blast or in natural storm is pinned and unable to escape.
- Can openers for canned foods.
- Local maps for use with information on safe areas for later evacuation.
- Prescription medications, eye glasses.
- Infant formula and diapers, if pertinent.
- Pet food and extra water for pets, if pertinent.
- Important family documents such as insurance policies, identifications, bank account records, in waterproof, portable container.
- Cash, or traveler's checks and change.
- Emergency reference book or information from ready.gov.
- Sleeping bag or warm blanket for each person. Additional bedding in cold climate.
- Complete changes of clothing and sturdy shoes, tailored to climate.
- Household chlorine bleach and medicine dropper. Disinfected water

is made with 9 parts water to 1 part bleach, or 16 drops household liquid bleach per gallon water. Do not use scented, color-safe, or bleaches with added cleaners.

- ➢ Fire extinguisher. But do not use with persons in closed space.
- ➢ Matches in a waterproof container.
- ➢ Feminine supplies and personal hygiene items.
- ➢ Soap, detergents, washcloths, towels, and large can for soiled items.
- ➢ Mess kits, paper cups, plates, plastic utensils, paper towels.
- ➢ Paper, pencils and lots of pens.
- ➢ Books, games, puzzles, or other activities for children. Books for adults.
- ➢ Inexpensive personal radiation dosimeters in everyone's pocket (**Exhibits 9 and 10**), and preferably at least one Geiger counter in each shelter (**Exhibits 11 and 12**).
- ➢ This Handbook, to review protections from radiation and when additional shielding might be needed (although not likely for the vast amount of areas outside the immediate blast zones). See following sections IVB and IVD and **Exhibit 6**.

Note: In addition to common items, it is desirable to have, for each family group, an inexpensive personal radiation dose measuring instrument in each person's pocket like those in **Exhibits 9 and 10**, and (if affordable) a portable geiger counter, like the ones shown in **Exhibits 11 and 12**. The chemical SIRAD dosimeters in **Exhibit 9** are very inexpensive (about $15 each), if ordered in quantity; the pocket ionization chamber in **Exhibit 10** can often be purchased for $100 or less, depending on its range, or some can be obtained in courses given to teachers by the Health Physics Society; the geiger counters in **Exhibit 11 and 12** could cost several hundred dollars or more each, but could be affordable if several persons or neighbors pitched in to buy a few. A list of vendors provided by the Health Physics Society (HPS) is in Brodsky (2011), and can be found currently by checking the HPS website.

IVB — BUILDING OR PLANNING TO USE SHELTERS IN HOME OR AT WORK

The planning or building of shelters can range from the simple makeshift shelter in a basement, as shown in **Exhibit 6**, using the information in **Exhibit 7**, to planning a place in a building using the protection factors shown in **Exhibit 8**. The kind of shelter built or planned at home will depend on the resources of the family. More livable shelter designs are included in my 2011 book, with reference to those in early recommendations of civil defense organizations in the 1950s-1960s. Shelters available commercially with installation can be examined on the site of KI4U. Materials should at least be stored that can be used to construct at least ad hoc shelters of the homemade type in **Exhibit 6**. Even the makeshift type of shelter shown in **Exhibit 6** could provide some likely protection from homes destroyed by hurricanes or tornadoes. There have been some reports on our weather channels of persons who survived tornadoes while most of their homes were destroyed, because they at least provided one strong room where they could survive.

EXHIBIT 6

Preparing a Makeshift Shelter with Planned Shielding Materials

PLANNING OR BUILDING SHELTERS AT HOME

The following **Exhibit 7** gives the density, and the thickness for $1/5^{th}$ reduction of gamma intensity, of commonly available materials that can be stocked or used at home to build shelters.

EXHIBIT 7

Table of Approximate Densities, and Thicknesses for One-Fifth Gamma Reduction, of Common Materials

(for Broad Beams of Radiation)

Material	Density (pounds/cubic foot)	Thickness for $1/5^{th}$ Reduction (inches)
Wood (birch, oak, maple, etc.)	40	17.5
Paper (compacted in books)	56	12
Water, or human tissue	62.4	11
Earth (loose)	75	9.3
Sand (dry)	100	7.0
Brick (common)	110	6.4
Concrete block (solid)	140	5.0

Solid concrete block is seen to be the most effective of the above materials for attenuating the gamma radiation from nuclear bomb fission products, and this would also apply to single radionuclides likely to be used in "dirty bombs." To use **Exhibit 7** for different wall thicknesses, note that a three-concrete-block thick wall, each block of 5 inches, would reduce the radiation intensity to $1/5 \times 1/5 \times 1/5 = 1/125^{th}$ of the outdoor intensity at ground level. Steel reinforcement using hollow concrete blocks filled with concrete after the steel rods are in place would also provide considerable protection against blast effects for the family.

Exhibit 18 in my 2011 book shows an above-ground shelter I designed and had built behind the doors to my patio in my home in the late 1950s. It had steel rods placed as shown between two concrete-block walls, with concrete poured into the blocks after the rods were in place, and then (after the concrete hardened) sand poured between the two walls in a space thick enough to give a radiation protection factor of $1/5000^{th}$ and a blast protection of 30 pounds per square inch max. This shelter was built in the cold war era to protect my family at five miles from a 15 megaton hydrogen bomb attack. Note that 15 megatons of TNT is equivalent to 1,000 times 15 kilotons (kT) of the Hiroshima and Nagasaki bombs, and equivalent to 15,000,000 x 2,000 = 30,000,000,000 (30 billon) pounds of TNT. My above ground shelter cost me $5,000 to build behind a $12,000 home in Baltimore County in the late 1950s. This kind of shelter would be the ultimate for protection against blast, radiation, chemical and biological agents (if properly sealed and ventilated through filters), and all types of hurricanes or other natural disasters above water levels. It would likely cost over $60,000 today, unless the government or some private consortium would promote it as a boost to the construction industry, with standard designs and materials.

PLANNING PROTECTION AT WORK

At work, a person should plan to seek the room in the strongest building in **Exhibit 8** that he can get to within a few minutes, without staying at or looking through any windows in the meantime. A blast wave from a nuclear explosion travels, after an initial burst, about a mile in each 5 seconds. A healthy person in an urban area can at least enter a nearby building in 30 seconds or seek an inside room of an office building. This action alone, going inside a building, with duck and cover as in **Exhibits 1 and 2**, will also greatly reduce any radiation dose or contamination from descending radioactive fallout.

Note that in **Exhibit 8** the deepest levels below ground can have protection factors > 200. Suppose there was 10 feet of earth alone between the underground shelter and the outside surface. From the table in **Exhibit 7** above, the 10 feet would be 120/9.3 = about 13 fifth-thicknesses. This thickness, even for loose earth with no protective concrete or building above, would give a reduction factor of about

$1/5 \times 1/5 \times 1/5 \times 1/5 \times 1/5 \times 1/5 \times 1/5 \times 1/5 \times 1/5 \times 1/5 \times 1/5 \times 1/5 \times 1/5 = 1/(62,500,000,000)$, which is less than one sixty-two-billionth. So, no fallout radiation intensity would remain in the lowest level in Exhibit 11.

A rule of thumb from civil defense days is that an area thickness of 300 pounds behind each square foot of wall would give a reduction factor to one five-thousandth of the incoming radiation. Thus, it may be assumed that the equivalent amount of weight behind each square foot of wall in my shelter above, of the late 1950s, would be equivalent to 300/(140 pounds per cubic foot) = 2.14 feet thickness of concrete. This is about right for the two-concrete-block walls filled in between with sand in my shelter of the late 1950s.

This information in the last two paragraphs can be helpful to anyone designing his own home shelter also. My calculations may be easily checked by careful use of simple arithmetic by the interested reader. Please let me know if I made any mistakes.

DISCUSSION ABOUT SHELTER POLICIES

As indicated in my 2011 book, early civil defense scientists designed and tested strong shelters that would withstand blast, as well as shield occupants from radiation. Switzerland passed a law in 1950 that every home must have such a strong shelter and it must be used regularly for some family activity so that it would be maintained in good living condition. About 95% of Swiss families have such shelters in their homes today. Other nations have followed suit.

It would be easy for our nation to put such shelters in almost every home, if someone in government or those in the building industry would take the lead. It would also help the construction industry. If we truly care about human life, let us urge our government leaders and builders to initiate national shelter construction. References to documents on shelter construction, a photo of an extremely strong shelter, and companies that can provide shelters, are included in Brodsky (2011) and available from Connor (2014).

It is strongly recommended that all citizens demand a better national shelter program as we urged in the early civil defense era of the 1960s, and as adopted by Switzerland in the 1950s. It would promote many important jobs for the building industry, rather just the overbuilding and selling of condominiums that have reduced home prices. More important, reinforced concrete shelters built to protect citizens from the blasts of atomic bombs at one mile or more distance would also protect, if properly sealed and provided with clean air for breathing, from many other toxic agents, and from other natural or other man-made disasters. A range of shelters for protection at different cost levels is presented in my 2011 book. Shelters are available in large, strong buildings, or underground, as illustrated in (Buddemeir and Dillon 2009; Buddemeir et al .2011), but the public has not been adequately prepared to use them. Purchases or assistance in shelter construction and emplacement may also be obtained from shane@ki4u.com , or by Googling the word "shelters."

EXHIBIT 8

Protection Factors at Various Locations in a Variety of Buildings

(Picture from Buddemeir and Dillon (2009) with adaptation of their information)

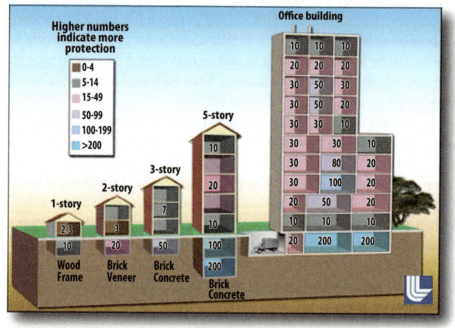

The protection factors in **Exhibit 8** range from low to high in order of the estimated approximate amount of protection from nuclear gamma radiation from deposited fallout after a nuclear attack. This means that a protection factor of >200 would reduce your exposure to **less than 1/200th** of that which you would receive standing on the open ground. Such a protection factor of 1/200th would likely save your life, even if you were in an area where the mushroom cloud deposited the greatest amount of fallout early after an explosion; of course, this would mean you were also in a protected location out of the 1-mile radius of heavy blast destruction, if the blast were from an approximately 15 kiloton bomb like those of Hiroshima or Nagasaki. The majority of an urban population would be farther away than that, but could still receive measurable radiation; the amounts of radiation would depend on wind directions and the presence or absence of rain or snow. These factors would also most likely approximate those from any radionuclides present in so-called "dirty bombs."

Note: Again. There is a less-than sign (<) before the 200 in the left indicator of number ranges. The deepest level below ground in a multi-story building will have factors much great than 200 (see previous page above indicating much greater protection in levels well below ground, as in an underground parking lot).

IVC — INSTRUMENTS OR DOSIMETERS TO MEASURE DOSES OR RADIOACTIVE CONTAMINATION

IVC-1. Personal wallet-size dosimeters you can wear for your immediate dose information. The most affordable and appropriate radiation dosimeter for emergency use that can be obtained for only $10-$15 in quantity is the SIRAD card-size, color-changing type shown

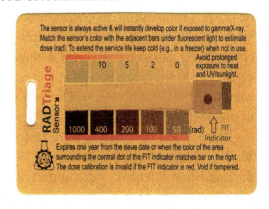

in **Exhibit 9** and **Exhibit 11**. The stamp-size SIRAD in Exhibit 11 can be stuck to a wallet or other ordinary pocket item. It could likely be obtained in quantity for only a few dollars, if some incentive was given to Dr. Patel by governments to re-introduce them for safety of the public.

This dosimeter is based on the fact that certain diacetylene polymers change color (or darkness) in proportion to the amount of radiation received. They are produced with a middle strip that increases in color (and darkness) only when exposure begins to enter a range **approaching** those where serious health effects might occur. Only when the middle strip is dark enough compared to a lower range of fixed colors of increasing darkness, are the doses likely to be harmful or result in fatalities. Thus, these dosimeters are easily read, do not require battery replacements, and do not scare those in the majority who are not likely to receive serious doses. They have been developed by Dr. Gordhan Patel, who has developed them over many years with millions of dollars of Federal support and considerable investments of his own. They have received many patents and awards, and are uniquely available only from Dr. Patel's company. He personally observes the quality assurance of every step in the manufacturing process for every dosimeter he manufactures.

(Dr. Patel is a fine individual and brilliant scientist dedicated to research and invention, as well as personal attention in developing the manufacturing of his own inventions, not only for his own profit, but also for benefits of society. I came to know him well after several visits to his company, where I wanted to examine his manufacturing processes before recommending his dosimeters. I had known of the SIRAD dosimeter research since the late 1990s, when a physicist who was then the project manager of his research for a Navy research laboratory gave me an early copy of the first SIRADs at a meeting of the Baltimore-Washington Chapter of the Health Physics Society. After carrying it for several years in my pocket, I was impressed with its stability and ruggedness.)

The SIRAD dosimeters were tested in USA laboratories soon after Marlow Stangler and I visited in 2005 with Dr. Charles E. McQueary, a distinguished scientist and engineer who was then an Under Secretary for Science and Technology in the Department of Homeland Security

(DHS) in the G.W. Bush administration. After field-testing with responders wearing them for their ruggedness, readability, and accuracy at the DHS top scientific laboratory in New York, the SIRAD dosimeters were approved by DHS for distribution to States and localities. Unfortunately, after Dr. McQueary was appointed to the Department of Defense and left the DHS, the DHS subsequently failed to recommend these dosimeters for purchase by the States for distribution to responders. Such distribution would have cost the taxpayer only an extremely small fraction of the billions spent by DHS on other equipment.

Such distribution of these dosimeters would provide information to each family on immediate needs for protection, and would avoid unnecessary fears at low exposure rates. They would have cost taxpayers an extremely small fraction of the billions expended on equipment provided to fire and police departments. Instead, a very limited quantity of much more expensive Geiger counters and gamma spectrum instruments were provided to only a limited number of responders in some major cities. These billions of dollars spent on equipment will be wasted if the public is in panic and floods the highways — preventing responder entrance, and exposing themselves to great dangers (see information on Fukushima in Cuttler 2013a,b).

(Reasons for the failure to distribute these instruments are covered in more detail in my 2011 book, with the hope that some legislators or administration officials might take notice and correct official culture. These reasons might have something to do with some current views that properly preparing civil defense measures for an attack might encourage enemies to attack us. This is like saying weakness is a discouragement for bullies to attack the innocent. Or, it is like reasoning that putting fire alarms in our homes is going to encourage us to start fires and burn our homes down.)

Many more reasons why I believe the wallet-size dosimeters pictured in Exhibit 9 should be distributed throughout the general public as well as to responders and medical institutions, in peacetime <u>before</u> unexpected releases, are presented in Chapter 8 of my 2011 book, and in the final appendix. They can provide persons wearing them, conveniently in pockets or wallets, immediate information about radiation exposures. Most persons will not see color changes (or darkness changes for the color-blind) and then they will know they are safe. The minority who might receive the higher, harmful, exposures will know that, when observing color changes or darkening, they need to seek better shelter and shielding, or at the several darkest levels to seek medical treatment as soon as possible.

I hope readers will obtain these SIRAD dosimeters before it is too late, rather than depend as in Fukushima on uncertain estimates of dose from uncertain instrument readings after damage and citizen deaths have needlessly occurred (see references by Cuttler 2013, 2014).

Consequently, I must appeal to the readers of this handbook to please insist that your States, local counties, or communities provide these dosimeters to the public at this time (see the urgency in Appendix C). At least, readers of this handbook should try to team up with others in the community or their governments, and buy some for their communities directly from the manufacturer.

I make this plea without receiving any monetary portion of sales.

Dr. Patel, the inventor, has recently sent to me his briefly stated reasons why everyone should carry an inexpensive credit-card-size SIRAD dosimeter in a pocket or pocketbook:

"A FEW MINUTES OF RADIATION TRAINING

Today's first responders are very busy with various responsibilities. Responding to a radiation emergency is just only one of them. They may face radiation incidents which may include nuclear bomb, dirty bomb, mishap at a nuclear reactor (e.g., Chernobyl, Ukraine and Fukushima, Japan) and a mishandling of a radiation source (e.g., Goiania, Brazil). Hence, we have prepared a few minutes of introductory radiation training for first responders (which can also be used by general public)."

Basic information: Avoid unnecessary exposure to ionizing radiation (e.g., gamma or X-rays); in large enough amounts they can cause cancer, injuries and death. Diagnostic dosages (chest X-rays = ~0.05 rad (0.5mSv) and CT scan = ~1 rad (10 mSv) are considered acceptable risks (except for fetus and children). Public is advised to limit their exposure to 5 rad (50 mSv) per year and 25 rad (250 mSv) for lifetime and emergency workers to 50 rad (500 mSv). There are no symptoms or medical treatment below ~50 rad (500 mSv) exposure. Contact an emergency room if exposed to dosages higher than 50 rad (500 mSv). Depending upon the dose and the dose period, nausea, vomiting and hair loss are usually the early symptoms after receiving radiation doses above 100 rad (1,000 mSv). If you are contaminated go to the nearest place and take a shower."

Nuclear bomb explosion: If you hear a huge explosion and see an extremely bright flash, drop and cover yourself for a few minutes. Keep eyes closed. Dangerous level of radioactive materials can fall (fallout) over a few tens of square miles. The fallout from the explosion looks like sand, ash or grit. Fallout loses 90% of its radioactivity every 7 hours, 99% in 2 days and 99.9% in 2 weeks, so stay indoors far from outside and behind heavy materials."

Dirty bomb/RDD: The major objectives of a radiological dispersion device (**RDD**) are to cause panic, worry and mass disruption. The radioactivity of the barely noticeable fallout is likely to be very low and limited to a few miles. While the area may be deemed unlivable, barely a few people may get doses higher than 5 rad (50 mSv). There is no need to panic."

Accident at a nuclear power plant/reactor: If it is a minor radiation leakage, it is extremely unlikely you will receive a harmful dose. If the accident is major/serious, e.g., a meltdown (as that at Chernobyl, Ukraine and Fukushima, Japan), the dose could be fatal (>1,000 rad)(10,000 mSv) for those who are near the reactor. Remain behind a thick object/wall or basement. Wait for instructions from the authority/government."

Improper handling of radiation sources: You will learn about these types of incidents (e.g., that of Goiania, Brazil) only after a handler is seriously injured. Once the incident becomes known, do not go near the affected area unless permitted by the authority. If you had been near the incident for a prolonged time, contact the authority. In case of a mishap with an X-ray or radiation therapy type machine, only the operator or patient may get over-exposed."

To minimize panic & worry purchase a radiation dosimeter: It is less likely that a radiation incident will occur and you will receive a dose higher than 5 rad (50 mSv). However, you would not know how much dose you have received without carrying a personal dosimeter. Accidents due to panic can cause more injuries and deaths than exposure to radiation."

Therefore, to minimize panic and worry, carry a dosimeter, e.g., wearable, instantly color-developing, pre-calibrated, always ready, reliable, rugged, federally funded and tested, field proven and affordable SIRAD˙ (RADTriage or RADSticker) for monitoring and triaging exposure information and treatment. RADSticker which weighs only 0.2 gram can be applied on many objects. A SIRAD may compliment, but cannot replace any other dosimeter/detector you may be required to use."

Final comment for this subsection: Dr. Patel's quote is consistent with my views above, except that his warning about shielding from reactor accidents would not apply to members of the public in the USA, who have not been, and will not be, near any failed reactor, like the responders who died trying to control the fire at Chernobyl in the Soviet Union. Also, Dr.Patel must get officials who still assume the linear-no-threshold (LNT) view of radiation effects to accept his statements. He can not be expected to speak about the hormetic or other research indications that show thresholds and no ill effects, and often positive health effects, at the lowest levels of acute or extended doses below about 20 rads (20 cGy) or

70 rad over one year as suggested by Cuttler (2014c). These effects were discussed above in Chapter II.

NOTICE: I have just been informed that Dr. Patel's SIRAD Dosimeters also qualify as accepted "personnel emergency radiation detectors (PERDS). This means they can be worn as a rad-triage badge 24/7, and that they actually have a useful life of several years if properly protected from sunlight. If there is a slight darkening of the center strip over many months, the reading at that darkness can be subtracted from the dose reading after entering a radiation area, to estimate the dose received during response to an event. (Personal communications Steve Jones, Rick Hansen, April 2014)

EXHIBIT 9

An Affordable, Simple, Personal Radiation Monitor

From the Award-Winning SIRAD® Family of Dosimeters

Available from:
JP Laboratories, Inc.,
ATTN: Dr. Gordhan Patel,
120 Wood Ave.,
Middlesex, NJ 08846.
Phone:(732) 469-6670,
 E-mail: sirad@jplabs.com

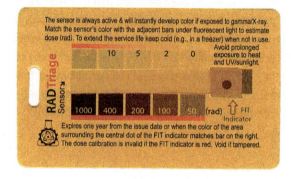

RADTriage-FIT™, a part of the **SIRAD®** (**S**elf-indicating **I**nstant **R**adiation **A**lert **D**osimeter) family of SMART dosimeters, gives you peace of mind that you are reliably monitoring your radiation exposure. RADTriage-FIT with amber filter provides significantly longer life under sunlight. It has a sensor with 2, 5, 10 and 25 rad reference bars above it and 50, 100, 200, 400 and 1,000 rad reference bars below it for triaging information in emergencies. The revolutionary **FIT™** indicator (on the right hand side of the sensor) simultaneously monitors false positives & negatives, overexposure to heat & UV/ sunlight, service-life and covers a portion of the sensor to monitor UV exposure. RADTriage-FIT is an affordable radiation dosimeter that is always active and ready to use, enabling disaster and emergency response personnel to measure their radiation exposure while dealing with the aftermath of a "dirty bomb" attack, nuclear explosion or an accident at a nuclear power plant. Batteries, calibration, and maintenance are unnecessary.

(More recent models of this dosimeter have been manufactured with dose scales in units of mSv, for use in Japan. If these had been distributed earlier by the Japanese government, so many millions would not need to be spent on estimating the Japanese population doses.)

IVC-2. POCKET IONIZATION CHAMBERS ("POCKET CHAMBERS") THAT CAN BE WORN ON A POCKET

Exhibit 10 shows the type of "pocket ionization chambers," sometimes called "pocket dosimeters," that were distributed to responders (fire and police departments, medics) throughout the nation by the Federal Civil Defense Administration and its successor organizations from the mid-1950s through about the mid-1990s. The Federal Emergency Management Agency (FEMA) unfortunately cancelled the program of providing nominal funds to the States to maintain instrument maintenance program, although these instruments had already been purchased with public funds. Only a few States continued to maintain this program on their own. The program also had provided geiger counters and higher range portable ionization chambers so that responders, or even citizens having access to these instruments, could have the ability to judge their personal doses and risks, and either avoid high risks or avoid panic at low dose ranges of insignificant or no risk, as explained in section II.

The cylindrical pocket chamber shown in **Exhibit 10** comes with a charger as shown on the right, which can develop a static charge and transfer it to the electroscope within the chamber.

Several ranges of pocket chambers are shown in the kit of **Exhibit 13**, with a charger that contains a battery, so that the chamber may be charged by holding it down firmly on a receptor, while looking through the lit-up cylindrical chamber until the line seen comes to zero (0), to start accumulating whatever radiation dose is received by the wearer after the charging is done.

A limited number of kits, or just the pocket chambers and a charger, can be purchased from Shane Connor's company at reasonable cost, although not at the price of $10 per chamber the U.S. government paid, when they the Federal government bought them by the millions and distributed them to the States.

The pocket ionization chamber is charged with electrons so the two parts of the thin wire loop repel each other, and are separated enough so that the magnifying glass inside the chamber shows a thin portion of the wire at zero when the chamber is held up to a light. As explained in section II, the amounts of exposure or dose to radiations we are interested in here, beta and gamma, are measured by the amount of ionization (electric charge) they produce in certain instruments designed to somehow immediately collect the charge, accumulate it with time, and cause a reading on a scale proportional to the dose that would be received by a person in the same location.

The pocket chambers shown in the Exhibits in this handbook would not measure beta radiation because the walls are too thick to allow betas to enter; they are designed for the proper sensitivity to gamma radiation for the range intended for each type of dosimeter. There were available at one time, special pocket chambers with very thin walls that were designed to be able to measure beta radiation. However, I have not seen them at any of the instrument displays at meetings for a long time.

EXHIBIT 10

Pocket Ionization Chamber and Static Electricity Charger

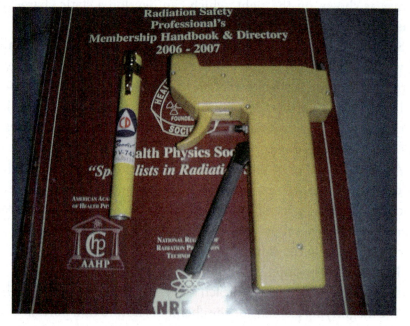

IVC-3. GEIGER COUNTERS OR GEIGER-MUELLER (GM) COUNTERS

Note: Some science organizations now recommend no capitalization of the word Geiger, even though it was a scientist's name, but we will not worry about that here. We will sometimes use the word with a capital G so you know it was the name of the original inventor. A scientist named Mueller improved on Geiger's invention, so these radiation detectors are also called Geiger-Mueller, or GM, counters. They have no relationship to the company General Motors.

Exhibit 11 and its accompanying subtitle visually demonstrates how a sensitive Geiger counter reading can be interpreted in terms of radiation dose rates that are easily measurable, but would not result in serious risks even after weeks of exposure. The small stamp-size SIRAD dosimeter shown in the exhibit might now be available only in limited quantities, because Federal, State, and local governments have not ordered sufficient supplies. Some can be obtained from KI4U.com, but the card-size SIRAD can still be obtained in large quantities from Dr. Gordhan Patel and is somewhat easier and more accurate to read.

A "two-piece" bench type Geiger–Müller counter with end-window detector.

Exhibit 12 is a recently-developed Geiger counter with a special circuit that allows measurement of exposure rates at background levels (10 µR/hour = 0.01 mR/hour) all the way up to 1,000 R/hour (= 1,000,000 mR/hour); the latter rate would produce, in a half-hour or less of exposure, a total dose that would cause death to more than 50% of those exposed within 8 weeks if they do not obtain special medical treatment, as indicated in **Exhibit 3**. The necessary kind of treatment is not likely to be available to many with the sparseness of qualified physicians in this specialty. Therefore, it is necessary to know when such high levels are present, even for a minority, although others not at risk must understand that the lower ranges measurable might not be dangerous depending on the amount of time exposed, and that the very lowest ranges would be of no health effects or even have some benefits to the immune system, as discussed in sections II and III.

The GM counter in **Exhibit 12** is able to cover such a wide range of intensities as a result of a special design by Phil Smith and his son Luke. The design is such that the instrument behaves like an ordinary Geiger counter until the frequency of pulses increases as intensity grows above a certain level where the pulse rate becomes so rapid that an ordinary GM counter could not recover the anode voltage for the next pulse. At that point, the Smiths' instrument activates a small, special computer and electronic circuit, incorporated in the encasement of the counter shown in Exhibit 12, that uses the time between pulses to calculate the pulse rate, and thus the intensity of the radiation field at the higher levels. I examined their factory processes before including their instrument in this handbook.

The Smiths have also matched their GM counter in **Exhibit 12** to special computer software and GPS equipment that allow continuous monitoring around cities or communities at specific locations. Because their GM instrument has such a wide range, it can be used to monitor gamma radiation levels at many specific locations, both for purposes of detecting the oncoming of lower exposure rates near background levels and/or monitoring exposure rates as they increase to levels requiring alarms to alert the public of the need for sheltering, evacuation (only when all circumstances dictate it appropriate), or additional shielding from gamma radiation. Their GM counter-computer system is now being located around the US, so that in the case any event releasing radioactive material occurs, it will be detected very early, and we will not be in the situation as at Fukushima, where radiation doses were unknown and inappropriate actions were taken that caused deaths when no chances of ill health were otherwise present. See the references by Cuttler (2013a,b; 2014a,b).

Geiger counters easily usable by any citizen, including the one in **Exhibit 12**, are also available from the KI4U site. Other instruments, including the one in **Exhibit 11**, are available from many other distributors listed on the HPS website or in my 2011 book. They can be purchased for prices in the range of $500 to $800, which is affordable to some families or groups of neighbors in many neighborhoods. These instruments come with simple instructions, which would be particularly meaningful to anyone learning the few words in Chapter II of this handbook or in Appendix A, and reading the information below.

**Those interested in a deeper understanding
may also now read the following paragraphs in this section.**

A Geiger counter, sometimes called a Geiger-Mueller (GM) counter, detects and estimates the intensity or flow rate through space of ionizing radiation such as gamma or x radiation. It has a tube at very high voltage that collects an avalanche of charge produced by the radiation in the chamber to create a pulse for each radiation interaction that knocks electrons out of the gas inside the chamber, or knocks electrons out of the chamber walls into the gas. This avalanche occurs before the knocked out electrons can recombine with the positive ions left behind, because of the very high voltage and the very thin wire, or wires, that collect each avalanche and produce a pulse.

The pulses are measured with a circuit that counts the individual pulses of avalanched charge. Thus, it measures exposure rate by measuring pulse rates, but the circuits can also cumulate the rate to give cumulated exposure (or dose) over time, as well as exposure (or dose) rate. It is designed with a fine wire "anode" that is placed at a the high positive voltage on a stable voltage plateau, typically at about 1,000 volts or more, with respect to a negative or ground voltage of zero at the wall surrounding the wire, which is insulated from the wall. Thus, the GM "probe" can be held without receiving a shock. All of the electronic circuitry that produces the voltage, measures the rate of pulses, converts pulse rates to exposure rates, and provides meter readings and clicking sounds, is contained in a rugged box, something like the yellow box of an early civil defense instrument in the upper left corner of **Exhibit 13**, except now there are smaller ones like those of the modern GM counters in **Exhibits 11 and 12**.

With a thick wall of metal (perhaps only in millimeters) beta radiation can not be detected, so the counter can detect only gamma radiation, unless there is a small open window of thin, protected, material that lets in beta rays when held close to the source (e.g., the skin if checking skin contamination).

Although most gamma rays of interest can penetrate the walls of a GM counter, enough gamma photons interact with the wall and release electrons into the chamber to produce the electrical pulses. These electrons are fast enough, so that they knock out electrons from atoms of the gas (or air) in the chamber with such energy that, under the high positive voltage attracting them to the very thin anode wire, they cause a huge avalanche of electrons that saturate the anode with a pulse of current. Each pulse is registered by an appropriate electronic circuit, which produces the clicks and meter reading, and then restores the proper voltage to the anode to allow recording of the next pulse. Restoration of voltage can occur within less than a millisecond interval so that thousands of pulses per minute can be recorded by an ordinary GM counter. The amplification in a geiger counter avalanche is such that about 100 billion (100.000.000.000) electrons arrive at the anode for each pulse that can be initiated by only one electron knocked out of an atom into the gas by a gamma ray. That is why a GM counter is so sensitive to radiation, even near low or background intensities that can do no harm.

If a very thin window is cut into the wall of the chamber and sealed with a thin plastic membrane protected by a screen, so that the gas in the chamber, usually at a pressure below atmospheric, can not escape or be infiltrated, then when the detector is placed near a surface contaminated with a beta emitter, the betas (fast electrons) can then immediately be detected very sensitively by the counter. The counter is connected to an electronic circuit and meter that indicates the number of counts per unit time, which, if no betas are entering the chamber, sometimes is roughly indicated on a scale to indicate the exposure rate of gamma radiation in mR per hour. The energy range of gamma radiation for which exposure is roughly indicated within a certain accuracy is specified in the information provided with the instrument. This accuracy of R or mR readings within an energy range is called "energy independence" in health-physics speak. Sometimes, the count-rate is also connected to an integrating circuit that can accumulate the counts as well as the indication of total exposure.

I hope that these explanations will give readers some understanding of how radiation detecting and measuring instruments work, because if they use them others will ask these kinds of questions.

IVC-4. ION CHAMBER SURVEY METERS

Ion chamber instruments, such as the older model yellow civil defense (CD) instrument in the upper left corner of **Exhibit 13**, are more useful at higher exposure rates. They are not as sensitive as GM counters, because they operate at voltage plateaus in about the 300-400 volt range, rather than 1,000 volt range, and do not have the extremely thin anode wires that produce extremely high electric fields. Therefore, they collect only the electrons at the rate produced inside the chamber by initial events, do not result in 100-billion-electron pulses for each event, and can indicate the steady currents produced by fields of radiation incident on the instruments' sensitive chambers. These ion chamber survey instruments are therefore not sensitive to levels near background radiation, and are more useful when expecting to be in fields where health effects should be limited. (When I jumped onto the island in 1954 at 30,000 mrad per hour, I took an ion chamber

Simple ion chamber radiological survey meter, Civil Defense, Model 715.

instrument (of size like the instrument in the upper left of Exhibit 13) with me to limit my time on the island; a GM counter would have been of no use, would have been saturated, and might not have had a reading at all. If an ion chamber instrument can be obtained inexpensively, it would likely suffice for informing the user about radiation intensities that would require additional protection and sheltering.

Comparing the size of the instrument in **Exhibit 13** with the GM counters in **Exhibits 11 and 12,** shows how miniaturization of GM counters now provides instruments that are more easily portable and yet can still cover wide ranges of radiation intensity.

IVC-5. OTHER TYPES OF RADIATION DETECTING AND MEASURING INSTRUMENTS

There are many other types of instruments for measuring the various types of ionizing radiation — gamma, x ray, neutron, alpha, and beta — but they are not important for the citizen just concerned with protecting himself and family in the kinds of terrorist attacks or accidents that might affect the general public. For incidents involving other kinds of radiation than gamma or beta, such as the poisoning in England of the Russian with alpha-emitting polonium-210, investigation of such incidents, or protection against them, requires the involvement of special scientists and agents. Also, anyone exposed to the prompt neutron radiation instantly emitted from a nuclear bomb will likely be in an area of total destruction, unless like the policeman in Hiroshima mentioned in the **Afterword**, he is well below ground in a protected location at the time of the explosion. So, neutron detectors are not of interest to citizens protecting against releases of radioactive material from either nuclear bombs or reactors.

EXHIBIT 11

Picture and Discussion of "High" Geiger Counter Readings and Chemical Color-Changing (SIRAD) Dosimeters vs. Possible Stay Times for Rescue or Seeking Shelter

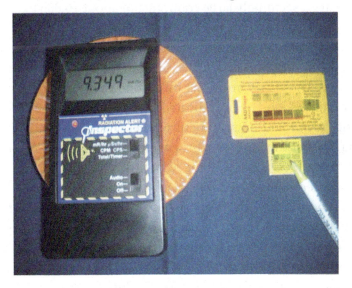

This is an exhibit of how I showed responders such as firemen and police how to understand that seemingly high Geiger counter readings do not necessarily mean danger in entering areas for rescue purposes. This GM counter is an affordable instrument available from Beth Cramer, S.E. International, Inc., P.O. Box 39. Summertown, TN 38483-0039, E Mail beth@seintl.com.

A counter like this should be available to every group of persons within a shelter in the event of a nuclear bomb detonation or other release of radioactive material. It comes with a small radioactive but safe source for checking its operation, and is also useful for instructions on familiarity with radiation phenomena. However, it is not in itself adequate for responders who might need to monitor their entry in radiation levels above 0.1 R per hour to save life. Note in Brodsky (2011) where I carried an ionization chamber monitor, rather than a GM counter, in a 30 R per hour field of gamma radiation to rescue my detectors after H-bomb tests, staying only about 5 minutes and then 20 on a contaminated helicopter. As a responder, I would gladly enter such an area to save any of you good readers.

After showing that the incident radiation from this uranium glazed ashtray gave a seemingly large count rate of about 16,000 beeps per minute, I then switched to the mR per hour scale to show what the exposure rate would mean in a gamma radiation field for the same 16,000 per minute count rate. The reading, as seen, is close to 10 mR per hour. The responders knew that m means milli (one thousandth as indicated in the definitions). So I then indicated that in a field of 10 mR per hour, it would take 1,000 hours (about 6 weeks) for a responder in such a gamma ray field to receive 10 roentgens exposure (10 R). This would be less than the 25 R allowed for responders for such life-saving efforts, with some leeway to receive even more. **Then, I say to the responder group I am talking with, "You routinely risk your lives staying in fires for as little as 10 seconds to save lives. When you could stay in such a field for 6 weeks, are you going to fear going in for 10-30 minutes or more to save lives? You would not even see a color change, or darkening, on the middle strip of this SIRAD dosimeter for a 10-30 minute exposure in such a field. In 1,000 hours it would not be up to 25 R." The responders then concur they would have no fear entering 10 mR per hour, which has been given them as a turnaround level in much official training.**

I then tell them they should be carrying all of the time such SIRAD dosimeters, which give dark scale indications from several R to hundreds of R, or at least have the smaller RADTriage badge pasted in their wallets. They would then know if they were approaching doses near or above the 25 R emergency limit. They face much more risk than that of 25 R every time they answer a fire or call for rescue than they would in the vast areas that would be in the mR per hour range, even after a nuclear blast.

Moreover, if they were indeed approaching higher levels of 30,000 mR per hour (which in my 2011 book I showed how I entered for five minutes to recover my detectors after the H-bomb tests of 1954) they would be alerted by an increasing darkening or color change of a SIRAD dosimeter within 5-10 minutes and could take measures to vacate the area or find adequate shielding.

Such instruction is necessary, because members of the public often hear statements such as, "Any radiation is dangerous," or "There is no safe level of radiation," even sometimes from officials or scientists who do not realize the needs for understanding levels of dose and risk under emergency conditions.

EXHIBIT 12

An Affordable Geiger Counter with an Extremely Wide Range Satisfying Requirements of Every Response Organization and Neighborhood

(See the paragraphs following this picture.)

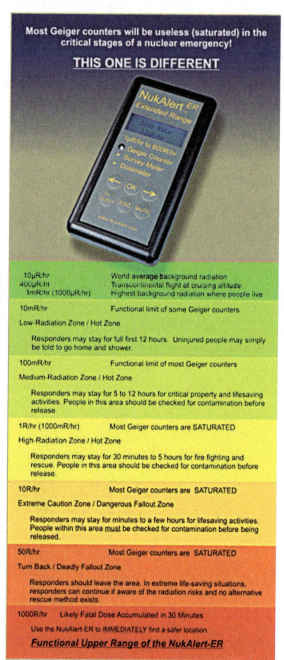

This instrument can be operated on two AA batteries and/or a USB power source (computer or universal cell charger). It is also ruggedized and will perform under a wide range of temperature and environmental conditions. Contact Shane Connor at shane@ki4u.com for price, availability, and specifications, and easy instructions for use. Marketed through Shane Connor as the Nukalert-ER, this radiation detector has recently been developed by Phillip Smith and son Luke, who have developed software that will also provide for a national and local radiation monitoring system using this GM tube. It can detect gamma radiation levels from natural background to more than 1000 R per hour. The NukAlert Data Gateway powers a NukAlert-ER Geiger counter and sends continuous radiation readings from it to multiple internet databases and an optional HDMI display. Power for the system may be supplied by a wall plug adapter, or power over Ethernet. Connection to the internet is by Ethernet or WiFi. The radiation intensity can be read as a BACnet variable and used to shut down building HVAC air intakes if a user defined radiation level is exceeded. The system is able to send live feed to databases such as the RadResponder Network. This system is already being installed at locations around the country, with GPS identification giving continuous monitoring of radiation levels at specific locations. This is more accurate and reliable than sending human surveyors into potentially radioactive areas in an emergency, where their radios and other equipment could be subject to failure or interferences.

This author hopes that, in addition to the Department of Homeland Security recognizing its importance, and placing such an instrument around the nation for central monitoring of radiation releases (before they occur, not like at Fukushima), it will also be available for monitoring by citizens and responders in every neighborhood.

EXHIBIT 13

Kit of Radiation Instruments and Information, with Ionization Chamber Survey Meter in Upper Left Corner

(Available from **www.KI4U.com**)

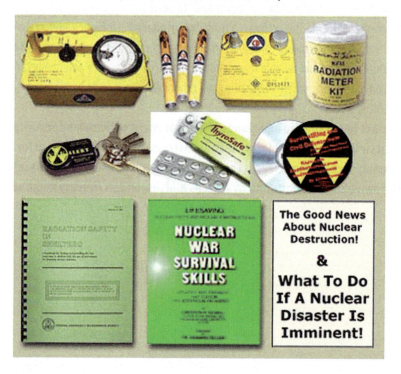

(Recommending this company is not favoritism. Shane is the only one who foresaw the need to collect these civil defense instruments and supplies in large quantities before they were destroyed in the mid-1990s, to provide them at reasonable cost for saving lives. You can find other suppliers of instruments from my list in my 2011 book,)

IVD — TRAINING FOR IMMEDIATE AID TO BLAST AND TRAUMA VICTIMS

Actions to prevent injury from blasts, or to attend immediately to help victims of trauma (Hatfill and Orient 2013), are applicable for natural disasters as well as any other sources of trauma. The importance of our learning the ways of aiding family members or others in our vicinity who have been traumatized or injured from blast or ballistic trauma are evident to us all. A summary of the lessons to be learned from the article by Hatfill and Orient are presented here. The original article should be consulted to prepare for the effective use of their guidance and instructions. Even better, a video recording and Power Point presentation of Dr. Hatfill's course, on which this article is based, is available from: ddponline.org/hatfill/. The most current guidelines can be found at C-TECC.org. Hatfill and Orient point out that anyone in the vicinity of a traumatized victim can provide life-saving aid.

Lessons from the Boston Marathon indicate that many healthcare professionals, as well as responders and ordinary citizens, could benefit from instruction on controlling catastrophic hemorrhage and removing airway obstruction. While Boston is relatively well-prepared in medical emergency facilities, even this city has many responders needing upgraded instruction in these emergency measures. Also, all involved need to be aware of the possibility of multiple explosions, and provide whatever protection available, if the trauma is the result of terrorist actions.

Priorities are different in mass casualty situations. In common medical emergencies the priorities are ABC: airway, breathing, circulation. In mass casualty situations, first priority is hemorrhage control: a person can bleed to death within 90 seconds to three minutes, whereas a patient can survive several minutes with an obstructed airway. Appropriate lightweight tourniquets should be available. Unconscious persons should be put on their side in the recovery position shown in the article. Bleeding should be stopped with tourniquets or occlusive dressings of open chest wounds. Triage will be necessary to avoid spending time on those who cannot be saved, at the expense of failing to save those who will live if given immediate attention.

Explosions can cause severe burns from thermal pulse, severe eye injuries from the blast and shrapnel and at least temporary blindness, hearing loss from rupture of the tympanic membrane by the blast, rupture of the tracheobronchial and alveolar airways and the gastrointestinal system from the blast and shrapnel, and brain injury and unconsciousness from the blast and shrapnel. All of these injuries can occur even beyond the range of complete destruction of an improvised nuclear device (IND) or an atomic bomb. Therefore, it should be remembered that a victim might not be able to hear, see, or feel any instructions during emergency aid.

Information provided by Hatfield and Orient (2013) for saving lives includes a very important section: Control of Hemorrhage, with pictures. The adult male has about 5 liters of blood; rapid loss of 2 liters will result in severe shock (a state of prolonged stagnated capillary blood flow to vital organs). An additional loss of 500 cc (one-half liter) will likely lead to death if not immediately corrected. Updated and some *ad hoc* methods of applying tourniquets are described, as well as packing body wounds.

Other information in their article includes:

➢ Airway management: If the victim is walking, he has a good airway. If the victim is unable to assist in recovery, methods of opening the airway and carefully placing him in a recovery position are described.

➢ Breathing: Check for open chest wounds but know you should not turn the person over before ruling out pelvic fractures. Sealing the wound is necessary; any material like plastic wrap or foil can be used. If possible, the casualty should sit up to increase abdomen pressure on the lung and make it easier to breathe. Other measures that probably require trained medics are described.

➢ Head-to-Toe Examination: This section advises checking for wounds on the front of the chest by running hands down the front of the body. Then, check for pelvic fracture by gently squeezing the body together. If no crepitus (crackling sound) is noted, then push posteriorly. The authors give further advice for checking injuries over the body, reassessing previous efforts at hemorrhage control, protecting injured eyes with rigid shields that do not put pressure on the globe, positioning the victim, and controlling the rate of fluid replacement to avoid diluting clotting factors. Assessing shock includes checking the radial pulse, which disappears around 80 mm Hg; the femoral pulse disappears around 70 mm, and the carotid pulse around 60 mm.

➢ Hypothermia: Blankets and thermal blankets are helpful, but active warming is likely to be needed such as using a hand warmer over the femoral artery or under the armpits, or placing the patient in sunlight.

➢ "The Blow-Out Kit": A kit is recommended to be available, but only trained responders are likely to understand what all the items are. The authors do, however, mention a number of ready items that might be available on your or the victim's body or in your car. Some of the information above will indicate the kind of ordinary items that might be kept in anyone's car in the event of the need for layperson's assistance to a trauma victim.

Important: This summary of the article is provided mainly to encourage professional responders to use the training in Dr. Hatfill's course, and encourage their supervisors to put on group courses. Their recommendations also encourage ordinary citizens to be familiar with much of the training materials, and have available the common everyday items that can be used to aid trauma victims. The training materials and contact with Dr. Hatfill for his courses can be obtained at ddponline.org/hatfill/ and access to this article and many others on emergency preparations can be obtained from jane@aapsonline.org.

Ranges of blast damage and trauma: Traumatic injuries, burns, and blindness can extend out to, and beyond the ranges of light damage to structures. Ranges and areas of severe, moderate, and light damage following a ground burst of three TNT energies of nuclear weapons are given in **Exhibit 14** below. The lower two are within the TNT range called "improvised nuclear devices (INDs)" in official reports, to avoid scaring the public about atomic (or nuclear) bombs. The lower yield weapons are believed to be of simple design and of easier assembly by small terrorist cells. Alvarez (2004) describes experiments that show that so-called dirty bombs can not significantly spread enough radioactive material to cause harm to the public, but that, "It is possible that a single individual could assemble a fissile device in less than a week that is transportable by small sedan." The "fissile device" he is describing is a small atomic (nuclear) bomb that could be as devastating as 1 kT, which is equivalent to 2,000,000 pounds of TNT. A 1 kT bomb requires the complete fission of only 56 grams of U-235 (Auxier 2004). This is only 56/454 pounds (less than one-eighth of a pound) of U-235, compared to 2 million pounds of TNT.

Although the ranges of light damage to structures, and possible traumatic damage to persons, can extend to over 13 square miles for a 1 kT weapon, and over 50 square miles for a 10 kT weapon, the ranges of fallout patterns that can make a Geiger counter click at hundreds of times the rate in natural background, can extend to many hundreds of square miles (Brodsky 2011). There are hundreds of fallout patterns shown in one of the references in my 2011 book, and none of them are identical. These patterns depend on wind speeds, their directions at various altitudes, their fluctuations, and precipitations and their locations. There really is no "typical" fallout pattern.

EXHIBIT 14

Ranges and Areas of Blast Effects from Nuclear Bombs*

(Adapted from University of Pittsburgh 2011)

TNT Equivalent*	Range of Severe Damage	Range of Moderate Damage	Range of Light Damage	Area of Severe Damage	Area of Moderate Damage	Area of Light Damage
10 kT	0.5 miles	1 mile	3-4 miles	0.79 square mile	3.2 square miles	50 square miles
1 kT	0.3 miles	0.5 mile	1-2 miles	0.28 square mile	0.79 square miles	13 square miles
0.1 kT	0.12 miles	0.25 mile	0.6-0.8 miles	0.045 square mile	0.063 square miles	2 square miles

➢ 1 kt = 1 kilton TNT = 1,000 tons of TNT = 2,000,000 (2 million) pounds of TNT. The ranges and areas of damage are approximate, and will depend upon buildings, trees, or other protective structures. The areas of measurable fallout radiation intensity will cover tens to hundreds of times the ranges in this table.

The bottom line is that:

➢ All responders and citizens hoping to be able to save traumatized individuals in their communities or vicinities should be updated on the best procedures for saving lives, by taking Dr. Hatfill's courses or using his video and Power Point presentations on ddponline.org/hatfill/, and the most current guidelines on C-TECC.org. Again, Hatfill and Orient point out that anyone in the vicinity of a traumatized victim can provide life-saving aid. This and other available articles on preparing for saving lives in emergencies can be obtained from jane@aapsonline.org.

➢ Responders and individuals hoping to aid trauma victims after a nuclear bomb or IND attack must not be afraid of the levels of contamination on victims that will be orders of magnitude lower when transferred by touch to their own skin or clothing, and can easily be removed later by simple washing. I know this from my own experience running around on a heavily contaminated island after the hydrogen bomb tests in 1954 on Enewetok, and from training and working with responders in real fallout fields at the Nevada test site in 1957 (see my 2011 book cartoons on this and the discussion surrounding them). Everyone should also be prepared to determine radiation intensities with simple, light, dosimeters or Geiger counters, so they will understand radiation risks vs. levels of radiation intensity, and not panic and avoid aid to the victims, over the large areas where neither they nor the victims will be seriously harmed by the radiation doses received. Radiation levels over most of the area where trauma victims need immediate care would not harm informed and trained responders, even if they stayed in the area for a number of hours.

IVE — TRAINING AND SUPPLIES FOR IMMEDIATE DECONTAMINATION AND WASTE REMOVAL

For the early weeks before some scientists or authorities might be able to descend upon you to attempt more sensitive detection of radioactive material in or on your body, you will be quite safe, likely forever, if you were not outdoors and in descending fallout, to just:

➢ Brush yourself off before entering your shelter, and wash your hands after inside, if you have been outside, or need to go out of the shelter for 10 or fifteen minutes to fetch some food products, other supply items listed in section IV-A that were not placed within the shelter in advance, or to use an outside improvised toilet.

➢ If you think you were exposed to descending fallout while outside, also take off your clothes and use soap and water on a washrag to go over exposed areas of your skin or hair, and dispose of any potentially contaminated rags or clothing in the special large waste can inside the shelter, lined with a plastic bag, within the shelter. The amount of radioactivity in the can will not be enough to cause an external radiation hazard. DO NOT SCRUB YOUR SKIN UNTIL IT IS RAW WITH OPEN WOUNDS.

➢ If you have obtained a Geiger-counter survey instrument with an open window for beta detection, of a type similar to those in **Exhibits 11 or 12**, and want to check me on these recommendations, note the following: On pages 122-125 of Brodsky (2011), I have shown it unlikely that while being outdoors beyond the lethal blast destruction areas, for either a nuclear bomb or RDD, unless you wiped your hand upon the ground, **your skin contamination would not read more than about 0.16 millirad per hour** on a Geiger counter with its beta window open and within about 1 centimeter of the skin. These Geiger counters will also measure any gamma radiation present in the area, with (for our purposes) about the same reading in mR/hour as a beta reading in mrad/hour. Thus, you would be safe in assuming this to be the maximum skin dose rate for emergency purposes.

Noting in **Exhibit 11** that a 10 mR/hour exposure rate, or the approximate corresponding whole body dose-rate, could be accepted by a responder for 6 weeks to still be under 10 R, where 25 R of acute (short-term) exposure is usually the suggested official limit for saving life. No ill effects, and no significant chance of late cancers, would be associated with an exposure of 10 R over a period of 6 weeks (see section II). **A count-rate corresponding to a reading of only 0.16 mR/hour, or even 10mR/hour (about 10,000 to 20,000 counts per minute) on a geiger counter with open window, would not have ill health effects, for either a reading of gamma radiation or beta radiation on skin.** This is true even though a reading of 0.16 mR/hour would give a count (click) rate about ten times that of natural background, and could be registered over a wide area beyond the range of blast or heat effects. (The beta dose rate for skin would be about the same as a gamma dose rate to whole body, for readings of an instrument sensitive and proportional to ionization density, for our purposes.) Also, such a dose rate indicated by a reading of 0.16 mR/hour is so low it would not even provide appreciable hormetic effects. (See section IIB.)

My views here are based on personal experiences on Enewetok Island with gamma radiation readings of 30,000 mR/hour from hydrogen bomb fallout and neutron activation of the soil, and from training responders at the Nevada Test Site in similar fields of radiation from fallout from nuclear bombs in the range of kilotons TNT, as well as the scientific literature and calculations referenced in Brodsky (2011).

V. How to Find Reliable Experts You Can Believe: Very Difficult But Doable

VA — THE BAD NEWS: THE DIFFICULTIES AND MY FAILURES

Difficulties: The uphill effort to reverse the growth of a vastly misinformed public: First, an apology and perspective: I apologize to journalists in the media about remarks in this book. I do believe that the vast majority of journalists working for various media are good and hard-working people who want to get true facts and important information to the public. They are also talented persons who are excellent at boiling information into limited articles to meet tight deadlines.

However, either because
- ➢ the hiring of scientists by these media is too expensive, or
- ➢ many journalists and editors are not aware of how to consult and hire the best experts in this field, or
- ➢ as explained later about Ray Johnson's findings at the end of this subsection, scientists generally are not of the personality to want to be journalists or appear in the media, articles in the media, and even debates on TV, are generally biased against a necessary true understanding of nuclear issues — those related either to bombs, nuclear power, or the many current uses of radioactive material that benefit industry and public health, and contribute hundreds of

billions to a healthy economy (Waltar 2004).

Similar false information has been fed to all the public in the USA for six or more decades, and not only unintentionally by well-meaning reporters who have no science backgrounds. This misinformation has also been provided by the entertainment industries that have exploited reports about the victims of Hiroshima and Nagasaki to pique the imaginations not only of adults in movies such as "On the Beach" and "The China Syndrome," but also our youngest children in cartoons and comic books that have attributed impossible health effects and genetic changes to radiation producing all kinds of monsters. Current (Cuttler 2014a and b) as well as earlier (e.g., Cravens 2010; Waltar 2004) examinations of scientific facts proves that this misleading of the public has produced unnecessary fear and death, while the radiation doses received could not have been harmful to anyone. Information in the entertainment media has also been a form of propaganda against nuclear energy, and prevented its environmental benefits, since the late 1940s.

When there is an attempted "fair" debate between views, usually selected is one scientist who is well-versed from his own research and publications recognized by his peers as important contributions, opposed by another with a scientific or medical degree but who is known by scientific peers to be expert in misapplying statistics to data to confound truth. This makes the public feel that there are only two extreme opinions of scientists in the field, that the matter is controversial, and that science has no answer.

In the same issue of Health Physics News containing the article by Edquist (2014), a sampling is presented of the enormous flow of misinformation in the three years since the tsunami caused the Fukushima Daichi nuclear power plant accident (Walchuk and Wahl 2014). The article also contains interviews suggesting how to combat this disinformation. Also, the article by Edquist shows a posted sign at a Japanese home where decontamination has been completed, the sign reading a "dose" rate of 0.22 µSv per hour. This level of dose rate would in the traditional units be given as 0.022 mR per hour, close to one of the lowest natural background levels to which the radiation levels on earth has decreased in the billions of years since creation. In such radiation levels, the human race has grown and thrived, except for its wars. Such a sign in Japan can only create fear of radiation in a place where the radiation level is so low that neither a harmful nor beneficial effect can occur. (See about my interview with a Japanese reporter after Fukushima further on.)

Although I have supported at times the right of the media to have freedom of reporting and not revealing sources, I have also recommended to some of my representatives that somehow we need to **specify how media rights need to be balanced with responsibilities for accuracy and truth. This would be difficult to legislate, but some insights on how media, as well as the public, can find reliable "experts" is presented at the conclusion of this section.**

My failures: This subsection shows how difficult it is for a scientist to get truth to the general public.

Despite my more than six decades in the profession of health and medical physics, my lectures to many students and a limited number of public audiences, talks at professional meetings on the needs and ways to get out and talk with the public about radiation, and spending thousands of dollars giving away my 2004 and 2011 books to my colleagues and public servants who might take a lead on spreading truth, I have failed to make a significant dent on the conscious understanding by the vast majority of the public of the facts about radiation health effects, benefits of radiation applications, and the need to prevent and prepare for the use of nuclear weapons against us. (See Appendix C.)

Failure 1: After the September 11, 2001 ("9/11") attacks, I thought it was an opportunity for those who understood radiation doses and effects to deliver a wake-up call to the public, and have them prepare measures for protection against terrorist attacks with radiation or other agents. I felt that we might inspire the government to re-institute the civil defense program that was destroyed in the mid-1990s as a presumed prize ("peace dividend") of ending the "cold war." (Think what is happening today.)

To me, the 9/11 attack was a new opportunity to interest the public in civil defense, in obtaining radiation detectors and learning about simple facts of radiation exposure and how to judge when radiation levels would be dangerous. I arranged a panel discussion for our local Baltimore-Washington Chapter of the HPS, in cooperation with related professional chapters involved with safety and health issues. The discussion also had panelists who were expert on effects of biological and chemical weapons and on the experiences with nuclear power plant accidents. The panel discussion was held in the Washington DC area in November 2001 and was relatively well-attended, with over 100 participants.

In December 2011, the then-President of the Health Physics Society (HPS), George Anastas, who had heard about the November Panel, called me and asked me to set up and chair an Ad Hoc Homeland Security Committee (HSC) for the Society. I suggested that he ask a younger member, who might have unlimited resources and liked to spend time thinking about nuclear destruction, to take on the challenge. I had wanted to work on this urgent issue on my own.

A few days later, he had not found such a younger member, and still wanted me to accept. Thinking "what I should do for my country," I accepted the job but for only one year. Within six months, I had all the reports and training materials to instruct responders and educate the public from my great subcommittee chairs, and submitted them to the Board of Directors of the HPS for their consideration, along with a program recommended to get us out to the public. I recommended a few simple changes in the ByLaws to make the HSC a Standing Committee of fifteen, and a program with procedures for cooperation of several committees and Board members to get all the members of the many chapters out regularly to talk with public groups and individuals, reporting back their activities at each chapter meeting, rather than just listening to speakers. I felt that chapter members, especially the younger ones, would feel a more important participating role, and chapters would also inspire greater participation and attract more members.

The Board did not accept any of my recommendations; they rejected them all, and continued an HSC for further activities and recommendations. Nevertheless, I kept trying after resigning as chair to give several talks on the subject to HPS members and members of the DC Section of the American Nuclear Society (ANS), and continued to approach several more Presidents of HPS and ANS about the matter, also giving some further courses, and assembling speakers and writers for the textbook for the 2004 Summer School, published as Brodsky, Goans, and Johnson (2004).

Possible Explanation: Ray Johnson, one of the editors of the 2004 book is an outstanding health physicist, and former President of the Society, who has a chapter about communication in the 2004 book. He has written many articles, presented many talks, and has a book to go with his presentations. He has made a lifetime co-profession, and has his own consulting business, dealing with the psychology and methods of communication. I have attended several of his courses since 1984, and taken his communication tests. He has found that scientists and professionals of the type in organizations like the HPS and ANS, when compared to most of the public, generally have quite different styles of receiving and delivering communication. His findings explain to me the reticence of most members of the relevant sciences, except the very few like my former colleague Dr. Sternglass, who has misled the public, to get out to talk with the public, despite the willingness of some to give innumerable talks at scientific meetings on the needs and methods of such communication. True, the HPS and ANS have done excellent work with Science Teachers Workshops and student science fairs. They also have programs to provide answers to persons who take the trouble to contact their websites. Unfortunately, however, these **passive** public information programs get information to only a small proportion of the public, compared to the growing misinformation that is getting to hundreds of millions, now through social as well as public media (Walchuk and Wahl 2014).

I must also plead guilty to being one of those who are reluctant to present talks to the public. However, I have forced myself more and more over the years to do so, knowing of the importance of getting scientific truth to benefit the public and their environment. I am still only 85, so I hope that when I grow up I will be like Ray Johnson.

And so, I failed to get the HPS to establish an <u>aggressive</u> attempt to reach the public.

Failure 2: Another example: I worked over eight years 1964-1972 in a radiation epidemiology study of nuclear energy workers, headed by Dr. Thomas Mancuso, M.D., as a co-investigator with Dr. Barkev S. Sanders. This study was sponsored by the Atomic Energy Commission (AEC) at the University of Pittsburgh when I was on the faculty of the Department of Radiation Health. Dr. Mancuso was solicited by the AEC as the lead investigator because he had experience uncovering ill health among workers in Ohio when he was the Ohio State Medical Officer. Dr. Sanders and I helped Dr. Mancuso with the data collection, radiation interpretation, statistical procedures, and much of the write-up of project reports. Dr. Mancuso admitted little understanding of statistical methods, even though he had taken elementary statistics in matriculating for the Master of Public Health (MPH) degree some years earlier.

Together, we developed a prospective radiation epidemiology study ("prospective-retrospective" using an almost total population of workers based on the enormous records kept in the atomic energy (AEC) program). After reporting at a Hanford symposium in 1971 my own work in collecting the radiation data and checking some of Dr. Sander's statistical analyses, and writing the part of our report on my own methodology and interim results, I left the University project in about 1972 to work full time as a physicist in radiation oncology at a local hospital, thinking the methodology and data for the project were in good shape. I was confident that my co-worker and project statistician, Dr. Barkev S. Sanders, would be able to help Dr. Mancuso, and would carry on further studies with different control populations and other worker populations, using his precise and exhaustive analyses.

But several years after I had left the project, Mancuso fired Sanders when Sanders wrote a large and detailed draft progress report that concluded there was no evidence of cancer caused by radiation in the Hanford worker population, having carefully made certain he had accounted for "healthy worker effects" by his own unique methods, using a large and appropriate control population..

After firing Dr. Sanders, Dr. Mancuso hired Dr. Alice Stewart, an anti-nuclear activist who had admitted to me after the Hanford meeting that she did not understand statistical methods, and they dismissed the careful prospective analyses prepared by Dr. Sanders and me. They used only the much less dependable case-control methods that Dr. Stewart had used in earlier studies, with some absurd assumptions about radiation dosimetry accuracy, and reported their presumed findings of radiation caused cancers at a health physics symposium in Schenectady in the late 1970s. Although Dr. Stewart worked with a mathematical statistician, Dr. Knealle usually, she directed the calculations he was to perform. A simple example of the absurdity of one of their claims is presented in an appendix of (Brodsky 1996).

I was shocked and dismayed by the loss of the careful work over the years, especially by that of Dr. Sanders. At the same symposium, when Mancuso and Stewart presented their presumed findings, more respected medical investigators refuted their report. Yet, the media headlined their report, ignoring the more reputable epidemiologists. Dr. Sanders and I published papers in the Health Physics journal refuting in detail why the Mancuso-Stewart so-called findings were improperly derived, with references to dozens of other reputable epidemiologists who also refuted the Mancuso-Stewart papers. Even after these papers were published, and after I testified later about this before House and Senate committees, and talked with reporters who were there but turned off the cameras after Mancuso testified, the media still reported the Mancuso-Stewart claims as truth. The Mancuso-Stewart articles still continued appearing in some "peer-reviewed" journals as well as the media, and this team became darlings of the environmentally-destructive, anti-nuclear-power, movement. (Much more detailed information on this issue, and the related published references, including testimonies before the House and Senate, are in Brodsky (1996, 2014).) Despite meetings with reporters, the media attention to Mancuso continued.

Thus, I failed again. I could not make the Mancuso and Stewart team understand statistics, nor could I get the media to report the refutations of mine and Dr. Sanders', nor convince the media to give adequate coverage of the many other competent investigators who also refuted Mancuso and Stewart. Many of these investigators who refuted Mancuso and Stewart were very prestigious internationally in the sciences of biostatistics and epidemiology. Some references to their work are in my books and published papers.

Failure 3: Another failure of mine was from 1961 to 1971, also when I was on the faculty of the Graduate School of Public Health, University of Pittsburgh. In about 1962, I was impressed with the application for a professorship in the Radiation Health program at GSPH by a Dr. Ernest Sternglass, who wanted to leave Westinghouse to pursue research on the development of digital x ray diagnostic imaging, which would greatly shorten exposure times and yet save and amplify images of good diagnostic quality but with much less radiation dose to the patient. I recommended to Dr. Niel Wald that I thought he would be a valuable addition to our faculty.

I soon found out, after Ernie was hired, that he had major side interests in opposing any civil defense protection of the public in the event of nuclear attacks, and also had a side interest in examining radioactive fallout from the then very-active atmospheric nuclear tests in Nevada and the Pacific. Ernie would often come to me with draft papers or back-of-the-envelope calculations of fallout doses and effects at various locations in the USA.

Soon after he joined GSPH, I thought he had some unique ways of looking at epidemiologic data as a physicist, and helped him obtain the assistance of a committee of biostatistics and epidemiology faculty to help him develop a proper research proposal for his hobby of radiation fallout studies. He had prepared a draft paper about some early fallout from nuclear tests in Nevada where rain had brought down radioactive fallout in substantial amounts in Troy, New York, a situation in which I thought there might well be some significant exposures to the public. However, one by one the other professors on the assembled committee refused further participation when they began to see his unreviewed claims published in the newspapers.

When his publication about the fallout in Troy, New York, was first turned down by a reputable journal, from then on he became a darling of the anti-nuclear-power movement, as well as against any civil defense life-saving efforts. His news reports and publications continued to be for decades a major inspiration to the anti-nuclear movements, since the 1960s. If interested in direct evidence of one of his most absurd newspaper claims, see Exhibit 3 in my 2011 book and my surrounding exposition of the true facts.

Exhibit 3 in Brodsky (2011) is an article by Sternglass published in newspapers around the country during the Three Mile Island (TMI) nuclear accident using my name, but misusing information in one of my publications. The article by Sternglass exaggerated radiation doses received by the public at TMI to be millions of times greater than was received. Sternglass was referring to information in one of my 1965 articles I had given him back then as a reprint, but in his TMI claim in 1979 of high exposures, either he had not read my article carefully or he was deliberately misusing it and misusing my name without referencing my article for anyone else to be able to find its content; he certainly was capable of understanding the article with his Ph.D. in engineering physics.

In my 1965 article, referenced in Brodsky (2011), I wrote it so that safer estimates could be made of effects from radioactivity releases than had been published in an earlier paper, before adequate study of the Hiroshima-Nagasaki human data was available. (Thus, anyone reading it will see I am a good guy trying to protect the public.) I had calculated maximum doses to the public from the complete release of most of the biologically important fission products that could escape from a nuclear power plant. With the data in my 1965 article, it would then be easy to estimate the effects of any release of fission products once the fraction released of that in the fuel was determined. The amount of each fission product in the total fuel of a nuclear power plant is well known. Amounts released can be estimated from the fractions that escape through multiple filters. Sternglass must have known that he was ignorant of the fractions released at the time he wrote his scary article and appeared on Saturday night TV, only two days after the beginning of the TMI meltdown.

These fractions were negligible at TMI for all the radionuclides contained in the reactor's nuclear fuel, except for the radioiodines, of which less than one-millionth was already known to be released at the time of Sternglass's article. Release of even the more volatile radionuclides that produce the major doses after a nuclear plant accident was very reduced by deposition on equipment and building surfaces and passing through high efficiency filters following a tortuous path into another building. The rare-gas xenons did escape but they do not react with body elements nor remain in the body long enough to give an appreciable dose (Brodsky 1982).

Yet, although my name was mentioned in his Associated Press articles published all over the U.S., I was not able to get any of my letters of refutation published. About seven of my colleagues saw the article in different locations in the USA and called me wondering why I was supporting Sternglass; I told them how I was absolutely opposed to the lie or misinformation he was propagating, but had no way to refute the AP press reports. **Thus, without effective access to refuting press misinformation, as in the Mancuso hearings, or in this Sternglass fiasco, my constitutional rights of free speech were essentially violated by press censorship.**

Any reader who is interested can check my facts for himself, by reading my 1965 paper, referenced in my 2011 book along with a 2001 publication checking it with later scientific data. The material is not "space science"; I explain in these peer-reviewed papers the numbers and simple facts that make my findings understandable.

More facts about TMI: When my friend Dr. Reginald Gotchy, of blessed memory, measured over 700 individuals from the local population around Three Mile Island in late 1979 soon after the accident, he detected no radioactive material within any of their bodies, even with the most sensitive detectors. This exonerated me, because I had advised my Division Director at the Nuclear Regulatory Commission (NRC) a few days earlier that the NRC should not order an evacuation but instead should advise the public to stay in place inside their homes or offices. When the NRC decided to order evacuations, as advised by others at the same meeting, they no longer included me in their response organizations.

That left me free, on a Saturday afternoon, only two days after the start of the TMI accident meltdown, to drive five hours to the home of my former student and friend, Ken Miller, who was then at the local Penn State University hospital, in charge of decontamination and safety management of any workers at the Three Mile Island plant who might be injured and come with radioactive contamination to the hospital. None showed up by the time I arrived, nor at later times. (This was not Chernobyl. Many in the public have asked, "What about Chernobyl?" when I tell them the types of nuclear power plants in the U.S.A. are safe.) I noticed, after arriving at Ken's home a few miles downwind from the TMI reactor, that a Geiger counter on the coffee table at Ken's home read only natural background intensities. Nevertheless, Sternglass appeared on major TV that Saturday eve when we were watching the news at 10 p.m., scaring people about the great dangers to those in surrounding Pennsylvania. Evidently, he had already submitted his lies (or misinformation — I cannot read minds) to the Associated Press.

On the Monday after the accident began on Thursday, Ken had me talk to the medical staff about the better actions of staying on duty within the hospital, but by then even some of the medical department chairmen had left their sick patients and evacuated the area with their families, as advised by the NRC to the local authorities.

In section VC, I summarize some information to explain why, for nuclear power plants as built in the USA, accidents, even those causing meltdown of fuel, do not expose the public to significant danger of radiation-induced health effects.

So again, another failure: I failed to tame Ernie Sternglass, even after ten years of effort. I have also failed over the years to get press attention to his and other's misinformation about the TMI and other nuclear incident information.

Failure 4: My inability to get radiation perspectives and recommendations for personal dosimeters to the Japanese, in an interview with a Japanese reporter, a couple of weeks after failure of the Fukushima nuclear plants.

Several weeks after the initial destruction of the Japanese reactors, an April 4, 2011, an article in The Washington Post (Englund 2011) indicated that most of the released fission products were deposited within the nuclear plant area. On April 5, 2011, an article (Higgins 2011) discussed the confusion existing over radiation levels and risks, and pointed out that a Japanese farmer within 15 miles of the reactors who refused to evacuate, and who with his family had been drinking their cows' milk, were found in final screening to have ingested no detectable radioactive material. The negligible risks were predicted at the end of Brodsky 2011) and have now been substantiated by the world's best scientists and physicians. (See also Cuttler 2013, 2014.)

In April 2011, a couple of weeks after the tsunami in Japan and the beginning of releases of radioactive material from the damaged nuclear power plants, a reporter from Japan, Mr. Kazumoto Ohno, came to my home with his assistant to interview me about possible radiation risks to the Japanese public. He had been referred to me because he learned from my department chair at the University of Pittsburgh that I had been a co-investigator with Dr. Mancuso (mentioned above). He took pictures of me, taped my answers and remarks, and his assistant took notes. In the end, I guess nothing mattered because I could not confirm for him, but had to deny as I had done before Congress and in my publications, the assertions by Mancuso and Stewart of findings of radiation-caused cancers in the Hanford worker population.. Thus, I suppose nothing good came out of my interview with the Japanese reporter who interviewed me in my home.

Failed again: I could not get to the Japanese public nor to their leaders through a Japanese reporter any more than I could after decades of trying to reach US citizens or leaders through the U.S. press. (I had already sent, in the first week after the Fukushima incident, a short information news release to many U.S. reporters to provide them with some understanding of radiation events, also asking them to contact me for a discussion. None replied, except one sent a "thank you".) It is unfortunate that Japan had since decided to close all nuclear power plants; that would be a disaster for the health and economy of Japan. However, there has been some hope lately that their leaders are starting to understand the facts.

Now, I will stop admitting my failures. They might soon be counteracting my successes.

In the next section, I will deal with the good news about how it is possible to turn around public opinions. Some of the suggestions I make in the next section are lessons learned from failures, such as it is very difficult to get the important truths to the public through news media. That is one of the lessons learned from failures. However, the next section will provide some examples, adapted more briefly from my 2011 book, about how most of our intelligent public can be turned around by direct communication — a lesson most dramatically expressed independently in the book by the originally-anti-nuclear Cravens (2010). There is no indication in her book that she had ever even heard about my efforts.

VB — THE GOOD NEWS: OPINIONS DETRIMENTAL TO HEALTH AND THE ENVIRONMENT CAN BE TURNED AROUND BY TRUTH; HOW TO FIND THE INFORMATION AND THE TRUE EXPERTS TO HELP

There are many pages in my 2011 book describing occasions where I could use plain language, or write papers, that did turn around some persons and audiences that originally had the wrong ideas about nuclear issues and radiation effects at the low levels the public is exposed to in the USA. A few examples from my 2011 book of how I found it not too difficult to turn somebody around are:

➢ The description under **Exhibit 11** shows how I briefed several hundred responders in the National Capitol Region who were being trained at an official DHS-sponsored course. I had been invited to sit in by the homeland security chair of my Baltimore-Washington Chapter of the HPS. The exhibit shows how I demonstrated that a GM counter reading of about 20,000 counts per minute, which they first were afraid was a high and dangerous intensity, corresponded to 10 mR per hour, in which they could stay for 6 weeks without exceeding the usual 25 rem suggested limit in emergencies. Provided with an overhead projector, I was able to make this demonstration in about 5 minutes, with answers to questions afterward. They enthusiastically accepted the free SIRAD dosimeters that Dr. Patel had provided to me. The attendees seemed to overwhelming respond positively to my demonstration. I was later requested to come back to further classes to repeat the demonstration. See the written material accompanying **Exhibit 9**.

➢ My participation writing a chapter in the book, *Nuclear Power: Both Sides*. The editor of the book and his young assistant, Dr. Michio Kaku and Jennifer Trainer, were both anti-nuclear activists. Bernie Cohen and I had been invited to write two chapters, but initially Bernie thought he would turn it down, because he knew the senior editor to be a biased anti-nuclear theoretical physicist and his assistant was influenced by him to be the same way. The young assistant, Jennifer, with a degree in journalism, was assigned to be my editor, and changed some material I had wanted in my chapter. However, both Bernie and I decided to complete our chapters because we felt the book could end up worse without them. At the very end, Jennifer, after spending months discussing the chapter with me, admitted freely to me that I had changed her opinion about nuclear power being dangerous. The reference to the book with Bernie's chapter and my Chapter, "Protecting the Public," is in my 2011 book.

> A talk that I gave at an evening panel presentation to The League of Women Voters in Cleveland, which was attended by an estimated 200 women from the large urban area. I was assigned by Robert Minogue, Director of the Office of Nuclear Regulatory Research, U.S. Nuclear Regulatory Commission (NRC), to present slides about how nuclear power plants were evaluated by the NRC, even though I was at the time working only on the materials licensing guides and projects. When a question was handed to me later, when my turn came on the panel, it was about nuclear waste safety. At that point, I stated it was not possible for me in a few minutes to refute all the misstatements made by the anti-nuke panel members, and that Dr. David Waite (who later became President of the Health Physics Society) was the expert on nuclear waste. I said, however, I would attempt my answer after showing a few remaining slides about my background, which had not been adequately introduced. (I had planted these slides because of the experience on a Saturday TV show in Miami after the TMI accident mentioned in my 2011 book in which I was introduced as a scientist from the government, debating the Nobel Prize recipient Dr. George Wald from Harvard). I asked for the next few slides, on which the first showed me kissing my sweet old mother on the forehead, while I said, "This is my dear mother; she taught me to always tell the truth, be honest and protect the family name"; the next slide showed a picture of my wife and children, and I said something like, "This is my wife and my beautiful young children; they eat the same food, drink the same water, and live in the same environment as your children, and I care just as much about environmental safety and public health as anyone else here."

After the meeting was over and I was waiting to go to coffee with my hosts, I was approached by the young woman activist who was brought by an anti-nuke high school teacher of middle age; both had been on the panel on "the other side." She said to me pretty much like this, "You are the first human being I have known coming from Washington!" I talked with her a few minutes and invited her to visit me at the Nuclear Regulatory Commission and I would show her around and explain the safety processes we use in licensing. She said she would like to come down, but a bit later I overheard her being scolded by the older teacher for talking with me. (Shades of KGB.) This experience demonstrates the need to show your audience, one person or 200, that you care, which is usually pointed out now in many presentations at professional meetings.

I will not summarize any further experiences covered in my 2011 book. The above should suffice to provide **the good news**:
It is possible for anyone with the facts to spread truth to almost anyone in the public with an open mind, simply by showing you care, introducing the facts in plain language, and talking to them directly in person.
It has now been pointed out in many presentations of the Health Physics Society (HPS) over the past few decades that you must show people you care about their feelings before they will care about what you say. The problem is: We need to get those with the facts who care out *en masse* to talk to the public. It can not be done through the media, or through official government agencies, so far. I still hope that whole armies of those who know the facts will talk directly to the public, as I attempted to do in my description of my **Failure 1 above**. I can still provide copies of the program I recommended to the Health Physics Society in 2002.

I have also pointed out some books by others in my references that would be helpful with presentations to the public, if a major portion of our population and political leaders would be encouraged to read them; they were written by great scientists and engineers whom I knew personally to be honest, dedicated to public health and welfare, and to be practically geniuses in putting scientific truths to words the public could understand — like the books of Drs. Bernard L. Cohen, Ted Rockwell, Alan Waltar, and Gwyneth Cravens, to name a few. If their books had been read by a majority of the American public or as best sellers, I might not have needed to write some of my own books, including this one. However, the distribution of the books by these outstanding professionals and writers were distributed to only about tens of thousands, compared to the hundreds of millions of USA citizens who can read. (Their books can sometimes be obtained at discounts by searching for them on Amazon.com.) I needed to be much briefer in this Handbook, as some of my friendly critics advised. Some did not want me to have any references, but I need somehow to document the supporting information I present. Certainly, only less than one person in 10,000 could have ever heard of me. Anyone picking up my book would look at it and put it down saying, "Allen who?"

Then, one morning in the last days of writing this book, I remembered the book by Gwyneth Cravens (Cravens 2010), the early edition of 2007 of which I had read about six years ago. She was a former novelist, writer of fiction and non-fiction in a number of magazines, and an editor at *The NewYorker* and *Harper's Magazine*. She had also been an activist in opposing a nuclear power plant in New York when she met a scientist, Dr. Rip Anderson at a friend's dinner party. Rip had been a scientist, engineer, and manager at a number of different plants in the nuclear energy production processes. He not only began to answer her questions about the safety or dangers of nuclear energy, but offered to escort her to see the various types of production facilities in the chain that ultimately produces nuclear energy and its by-products. Following her education by Rip, she changed her views about radiation risks and nuclear energy, took on the challenge of writing a book, reviewed much of the published literature on the subject, and interviewed many of the scientists and engineers whom I have known and who have been leaders in nuclear science and engineering. A few of those have written books of their own that should have been read by more citizens. I have found Cravens' book so interesting that it is difficult for me to put it down. The serious communicator should also read the books in the references by Bernard L. Cohen, Theodore Rockwell, and Alan Waltar. Cravens has also referenced their work, as well as many other competent scientists and engineers.

Therefore, my job in writing this section has been made easy. **The ways you the reader can find the competent experts to learn the truth, and spread the truth**, about the safety of nuclear energy, and check the facts about radiation effects in this handbook, include:

➢ Obtain a copy of *Power to Save the World* by Gwyneth Cravens (Cravens 2010), read it, mark up what you feel are important sentences; and discuss it with, and recommend the book to, your friends, and especially to your congressional representatives and staff. Also, get members of the health physics and nuclear engineering professions to use the information in Cravens and in previous books by some of our best scientists and engineers to prepare and give direct talks and information to their families, friends and communities. See especially, for plenty of good facts and information, Rockwell (2003), Waltar (2004), and Cohen (1992, 1983); and also as references for the general citizen, Hiserodt ((2005) and Goldberg (2009).

➢ Invite a member of the American Nuclear Society or Health Physics Society to speak to one of your local organizations, or contact the **SARI** organization below for a recommendation;

➢ Check with the public information sites of the HPS and ANS, and their Q and A services; call their officers and contacts on their web sites. Many in the these professional organizations have been dealing with methods of estimating risk using the LNT hypothesis, and might not be familiar with radiation protective or hormetic effects at low doses, mentioned earlier in this handbook. However, they will be able to speak to the absurdities of those who have grossly exaggerated radiation effects in order to defeat nuclear power and civil defense efforts, even if the LNT were applicable as an upper limit to long-term risks, (see Walchuk and Wahl (2014) at http://www.hps.org). In the March 2014 issue of Health Physics News, Walchuk and Wahl describe the enormous flow of misinformation in the three years since the tsunami caused the Fukushima Daichi nuclear power plant accident. The article also contains interviews suggesting how to combat this disinformation. In this same issue, Edquist (2014) shows a posted sign at a Japanese home where decontamination has been completed, the sign reading a "dose" rate of 0.22 μSv per hour. Such signs scare people about radiation, even when it is at harmless levels. This level of dose rate would in the traditional units be given as an exposure rate of 0.022 mR per hour. Such a low rate is close to one of the lowest natural background levels to which the radiation levels on earth have decreased in the billions of years since creation. In such radiation levels, the human race has grown and thrived, except for its wars. The continued emphasis by the media on **radiation** after the Fukushima accident does not provide a true perspective on the negligible risks of these radiation exposures to the Japanese public.

> In Japan, consult groups of scientists (physicians, physicists, health physicists, and radiobiologists) who are associated with the leading academic institutions and the reliable Radiation Effects Research Foundation (RERF) studying the Hiroshima and Nagasaki populations exposed to atomic bomb radiation;

> In the USA, consult scientists affiliated with a number of academic and research institutes (e.g., check the website of the National Council for Radiation Protection and Measurement (NCRP). However, these semi-official expert groups often have members who are among the best researchers or doctors, yet have some bias against discussing hormetic effects at low doses and dose rates. We have a Congress that does not understand that low-dose research with radiation can lead to discoveries that help find cures for cancer and other diseases. I suspect there is some legitimate fear among members of an organization like the NCRP that if they propose publicly that low radiation doses are not dangerous, Congress might well reduce all low-dose research funds to zero. We need more scientists and top medical researchers in Congress, but they are usually of the personalities found by Ray Johnson who would avoid public office, although that is somehow not so in France (Brodsky 2010);

> To obtain the best scientific information on the facts about protective effects of the human immune and related systems at low radiation intensities, and the beneficial effects of certain applications at high or low doses, contact a new organization of some of the top scientists who have investigated this subject at the website of **Scientists for Accurate Radiation Information (SARI)** at http://radiationeffects.org. (Note: Radiation-Effects.com is a different site.) The articles by Cuttler (2013, 2014) are among the sources on this site, and many more lead to publications and sources of information on this subject. See the statement by the founder of NCRP, Dr. Lauriston Taylor (Taylor 1980). Lauriston Taylor, of blessed memory, once told me how many times he tried to educate the media over many decades, without much success.

Also, members of the **Association of American Physicians and Surgeons** can provide information on the levels of radiation exposure that are below the range where harmful effects would occur. This association can be reached at the website www.jpands.org. Also, trust those scientists referenced in my publications who have performed the basic radiation research, and who are listed as trustworthy in my references.

I must opine that Gwyneth Cravens, whom I have not met, must be a genius or extremely brilliant, as well as a great writer, to have digested such vast material from her many interviews with top scientists and engineers, and from reading scientific books and articles, and to have produced such a comprehensive treatise on these issues, with enough references to document and support all of her now-adopted views about nuclear power and its by-products. Her book alone could well have the "power to save the world," as the title implies.

Although Cravens' book sold well for this type of book, she informed me that it sold about 30,000 copies. With a USA population of about 100,000,000 readers, her distribution amounted to only 30,000/100,000,000 of the population, which is a fraction of 0.0003, or 0.03% of our population — not enough to make a political difference compared to the millions who have been misled by the information generally available to the public from entertainment and media sources. This is also true of the other books by outstanding scientists and engineers whom I have referenced.

Bottom Lines:

- If we can not get our better scientists and engineers out *en masse* (in the thousands) to talk with the public, those of us who understand the problem must get Craven's book in the hands of many who might have a bully pulpit from frequent appearances on the news or C-Span, especially our congressional representatives who are the best presenters and who can also influence national policy. The book is rather inexpensive and can be presented to our representatives and staffs as dedicated gifts.

- Decades of effort by some of our leading scientists, including Dr. Lauriston Taylor, who founded the NCRP, have not been able to get the media to provide an accurate picture of the safety of nuclear energy and its by-products, as regulated in the USA for decades based upon NCRP standards. Dr. Taylor, who passed away at age 102 some years ago, once told me how many times he approached the media to correct misinformation and was ignored. (Thus, I should not feel so bad.)

- Perhaps it is not likely after so many years that we will get scientists and engineers out *en masse* to present truths and correct falsehoods. Very few are like Ray Johnson on one side, and Ernie Sternglass on the other — both persons I have known well. Perhaps it because our studies required so much burial in books and calculations that we had little time to develop social skills, and perhaps that is why many of us are called "nerds."

- Decades of effort by many of us proves that we will not get the job done through the mass media, nor through government agencies' public information systems, which must always be conservative and politically correct.

- Without turning around the public and our representatives (who come from the public), we will continue to burn coal, produce air pollution, and destroy the environment digging it up, continue to blow up homes with failed gas lines, and continue to spill oil on pristine water resources and farms. The so-called "green" sources of energy will never provide us with economic and public health. Just read the book by Cravens and those by others I have referenced. — Ted Rockwell, Alan Waltar, and Bernard L. Cohen. I have known and admired all three of the latter. I admire Gwyneth Cravens whenever I pick up her book.

> France got it right. France adopted our light water reactor designs because they were more efficient and safer than their own under deGaulle's administration, "Frenchified" them with a few changes to retain their national pride, got government and industrial factions together to provide standard, approved designs, and now have about 80% of their energy produced by pollution-free, carbon dioxide-free, and environmental-destruction-free nuclear energy. Some of their energy is used to help nearby nations. However, France is an old country and has had much time to learn from some of its many mistakes. The USA is much younger, so in the matter of energy production, perhaps there is hope that the USA will be more like France when it grows up (I am dealing here only with the matter of energy production). Please read the book by the history professor who studied how France politically and socially was able to advance nuclear power, or at least read my two-page review of the book (Brodsky 2010).

VC — FACTORS THAT LIMIT THE RISKS FROM NUCLEAR POWER PLANTS WITH MOLTEN NUCLEAR FUEL

Experiments by Beard, summarized in (Brodsky and Beard 1960), were performed in 1957, in which two nuclear fuel elements, each containing thousands of curies of fission products, were taken bare — outside of all of the protective barriers that are in an actual nuclear power plant — and melted down in the Idaho desert, surrounded by the most inflammable materials and fuel that might be present in an aircraft crash.

Results showed that for the refractory material of which nuclear fuel and its coverings are made, almost all of the nuclear material except volatile gases remained in areas the size of large rooms. Although some radionuclides could be measured hundreds of meters downwind (by very sensitive instruments such as Geiger-counters), the vast amount of area where these radioactive materials could be measured was not at radiation levels that would be dangerous to health.

The field experiments described briefly above, and many others since, explain in part the insignificant health effects of the limited releases from the Three Mile Island (TMI) accident of 1979 (Brodsky 1982). Media reports to the contrary, which frightened many in the public to take improper actions and harbor unfounded fears even to this day, were taken from interviews of completely uninformed scientists. Again, you may examine in an e-book version of my 2011 book, in Exhibit 3, my exposition of the truth about the releases at TMI, and see how it documents how Dr. Sternglass falsely accused the government, particularly the Nuclear Regulatory Commission, in an Associate Press article all over the nation, of lying about radiation releases only two days after beginning of meltdown. In my 2004 and 2011 books, I provide tables of the quantities of biologically-significant radionuclides present in nuclear reactors of various operating and cooling times, as well as the amounts in short bursts (atomic bombs). Brodsky (1982) provides the fractions of radionuclides released at Three Mile Island, as determined some weeks after the TMI accident.

Thus, Sternglass was the one who lied to the public (purposely or not), it was NOT the NRC that lied, and he used my name in the article twice without my seeing his article beforehand, and without any way to refute it in the public media, only in my articles and books. That is why I must repeat it here. This is just one example of the many I have personally experienced of misinformation about radiation given to the public that I document in my 2011 book). (Also see my other book references.)

As mentioned earlier, not only were the doses to the public that I had predicted at TMI negligible, my colleague of blessed memory, Dr. Reginald Gotchy, placed in a very sensitive "whole body counter" over 700 persons who were in the vicinity of the TMI releases. He found no trace of radioactive material in anyone's body. The Pennsylvania health department radioepidemiological studies found no excess cancers in the population after the TMI accident. Beard's field burning of fuel elements in pools of aircraft fuel surrounded by combustibles, mentioned above, as well as later studies of fractional releases and inhalation literature, had much to do with my predictions at the end of my 2011 book about Fukushima releases. These predictions were substantiated very early by Englund (2011), and in the ensuing years at meetings of the Health Physics Society and the NCRP. The primarily local deposition in Japan, combined with the great dilution of any radioactive plumes and their fallout as they passed into the vast Pacific Ocean, resulted in no possibility of health effects in the USA. Brown (2011), quoting several government officials soon after the Fukushima disaster, was one reporter who was on the mark.

Why many actions to protect against harmful radiation levels and blast of bombs also protect against other agents and natural disasters: The actions recommended earlier in sections IVA, IVB, and IVD, to stock emergency supplies, provide available shelter at home or work, and prepare to help trauma victims, obviously can be helpful in saving life not only from events releasing radioactive materials, but also from many other kinds of natural and man-made disasters:

Measures recommended to avoid inhaling or ingesting radioactive material, or staying away from passing clouds that could contain toxic materials, would evidently save lives in the event of some natural disasters or events such as those in 9/11, where many responders have later suffered from breathing the debris from the falling buildings, which evidently contained asbestos as well as other toxic dusts and fumes. Thus, preparations to save life in the event of nuclear bomb or RDD attacks would also protect against other types of disasters.

VI. Conclusions

As news reports are beginning to show, there is little likelihood that members of the public in Japan, or even for any of those within the fifty mile evacuation area recommended for United States citizens, will receive radiation doses that will cause early or late harmful effects. This is partly because radiation exposures received internally will occur over time, be relatively low, and be compensated by natural repair processes (just as very limited exposures to the same amount of sun, very gradual steps of tanning, do not cause serious burns or the higher likelihood of cancer). Chapters 1 and 2 of (Brodsky 2011) show the exaggerations of radiation risks in much of media presentations, and Chapter 4 includes examples of hormetic effects of low, extended, exposures to radiation (see also Cuttler's references).

Those workers assigned to nuclear plant recoveries, however, could sustain serious health effects or even die if the officials do not enforce well-known safety provisions and emergency limits of exposure, whereas the public could panic and cause more deaths than would occur by radiation at lower exposure levels without adequate knowledge and preparation in the event of a release of radioactive material. As Cuttler has shown, 1100 persons died in Japan unnecessarily, and many more suffered away from their homes, because of restrictions improperly proposed by authorities who were not adequately prepared to differentiate between risks or lack thereof at various exposure levels. I do not believe that the Japanese officials and plant managers are not competent. They just did not adequately prepare the public, local responders, and the press to avoid panic, and recommend only needed evacuations.

There are some who have been influenced to oppose nuclear power in the USA who have been distributing exaggerated fears of all nuclear plants; it is my prerogative here to caution that these persons do not have adequate backgrounds to support their fears. The latest nuclear facilities proposed have additional safety features that would prevent the occurrence of the type of situation currently in Japan, even for the worst types of earthquakes. It may be noted that France has seen it fit to adopt our pressurized water reactors in the 1970s and now has almost eighty percent of its base electric power obtained from nuclear energy (Brodsky 2010). A reading of the book by Gwyneth Cravens (Cravens 2010) should be considered essential for anyone interested in understanding the safety and environmentally friendly aspects of nuclear energy, compared to the environmental devastation continuing to occur with the necessary use of carbon burning if depending on expected development of the inherently limited "green" energy sources other than nuclear energy. Cravens, who turned away from being anti-nuclear energy after examining all aspects with an open mind, also deals with the questions of the health effects of radiation, and also their absence in the current safety of applications of nuclear energy and its many beneficial by-products in the USA.

The facts in this short handbook are supported by further information and documentation in my 2011 book. A few selected references from the 2011 book are included in this book to support information and recommendations in this book; a few additional references are also included. My opinions also result from over sixty-five years of personal experience. See **About the Author** on the last page of this book.

This handbook also provides the minimum information needed for understanding and placing in perspective the news reports about releases of radioactive material.

Hormetic effects are not yet recognized in government regulations, which are designed to keep levels in peacetime very low to prevent larger exposures. However, when radioactive material is released unexpectedly in large amounts, the situation is not then under regulatory control, so it should be apparent that the real effects, non-effects, or even hormetic effects, must be taken into account in planning responder or remedial measures. Otherwise, more harm than good can be done by inappropriate actions such as mass evacuation from areas where more deaths than lives saved could occur, as at Fukushima.

Only a short summary of health effects is presented in this book. More is detailed, with references, in my 1996, 2004, and 2011 books. Shane Connor's **Afterword** (Connor 2014)), with "tongue in cheek," will suffice to succinctly make the point that the vast majority of us, even some homeland security leaders, are grossly uninformed about ways of surviving radiological dispersal. Decades of exaggeration of radiation effects in the media have caused generations of citizens to fear even the smallest radiation exposures, below even levels that would not entail any risks of harm.

Again, actions for surviving nuclear blasts and radioactive fallout would also do much to improve survival from other blasts or severe natural disasters, as well as from the release of other toxic agents. Some of the simple actions in the exhibits should be posted for frequent reading and reference.

May this handbook induce many to prepare for saving lives in the event of man-made or natural disasters, and to demand from our governments a greater effort to provide shelters, and empowerment to its citizens with dosimeters and information, before such disasters occur.

APPENDIX A

Glossary

Note: Many of the words defined in this glossary are not necessary for understanding this book, or the recommended preparations and protections for survival. They are included for those readers who might want to read some of the scientific documents or books listed in the References that support statements or facts in this book.

Absorbed Dose — The amount of energy absorbed in unit mass of matter, such as human tissue. It may be expressed in units of ergs per gram, rad (= 100 ergs per gram) or joules per kilogram (J/kg), where the latter has been named the gray (Gy) in the more recent Systeme Internationale (SI) units.

Activity — When those in the profession of radiation protection or nuclear science use this common word, they are often using it as a contraction for the word "radioactivity." Refer to the definition of radioactivity.

Acute Dose — An acute dose is a radiation dose received over a short time, usually a day or less.

Alpha Particles — An alpha particle is a particle ejected from the nucleus, mainly for radioactive nuclides of very heavy elements. It is composed of two protons and two neutrons, and is the same as the nucleus of the helium atom without its two outer electrons.

Atomic bomb (or Nuclear Bomb) — Usually refers to a bomb like the ones at Hiroshima or Nagasaki, where very **heavy atoms like uranium-235 or plutonium-239** break in two (**fission**) and a few more neutrons are released that break up more and cause an avalanche of energy release, according to Einstein's $E = mc^2$ because the total mass of the two fission products and the released neutrons is less than the original fissionable atom. When the energy avalanche has reached "criticality" it increases exponentially in small fractions of a second, and if contained in a heavy steel case it becomes a bomb and detonates when its heat and pressure break the case; when the case breaks, an explosion occurs and a strong pressure wave (blast) is propagated through distances in the surrounding atmosphere.

Background Radiation — Natural sources of radiation from materials in the ground, cosmic rays, or natural constituents in the human body such as the K-40 in potassium.. Sometimes the term is used for any radiation that interferes with the radiation one is attempting to measure at the moment. In such cases, the small amounts of cesium-137 and strontium-90 in the environment or in the human body from the worldwide fallout in the era of atmospheric testing of nuclear weapons, are considered background radiation when trying to determine more recent additions to environmental radiation. See Chapter III for examples of background radiation.

Becquerel (Bq) — The Becquerel (Bq) is a unit used in expressing a quantity of radioactivity. This name by itself is not capitalized in the scientific literature when it stands alone, even though it was the name of an earlier discoverer of radioactivity. One Bq is the quantity of radioactive transformation of 1 transformation per second (1 tps); it is sometimes called "one disintegration per second (dps)."

Betas (or Beta Rays or Beta Particles) — A beta is a high speed particle, identical to an electron, that is emitted from the nucleus of an atom.

Bioassay — As used in discussing the detection or measurement of radioactive materials taken into the body to assess doses delivered to various organs or tissues from radionuclides concentrated in different tissues, it has a different connotation than the same word used in other fields of biology. Here, it refers to determining the concentration of specific nuclides or mixtures in the various organs of the body, and the evaluation of their radiation doses over time. In radiation protection work, the term does not mean the assessment of changes in biological markers, symptoms, biochemical or physiological changes, or health effects. Such biological effects are assessed separately based on the specific factors involved — the specific type(s) and amount(s) in the various organs of interest are used, however, to assess radiation doses and risks. Also, methods of assessing doses, risks, and effects are also included in annual "bioassay" meetings of scientists active in this field.

Biological Dosimetry — Biological dosimetry is a branch of the field of dosimetry that uses biological samples, usually taken from individuals who have been exposed to radioactive materials or radiation, as a means to assess a high dose of radiation that could be at levels harmful to life or health. For example, biological sampling of hair, nails, or blood can be used, with different laboratory tests that depend on the type of radiation involved (e.g., neutrons, gamma radiation, or a mixed exposure). This kind of (biological) dosimetry is usually only sensitive enough to determine whole body exposures in the higher ranges.

Biological Half-Life — The biological half-life is the time taken for half of a radioactive material (which has been taken into the body by ingestion, inhalation, or injection) to be removed from a specific organ or tissue by physiological processes. Even for the same radionuclide and the same organ or tissue, the biological half-life depends on the biochemical form the molecule to which it is attached takes on when absorbed by the specific organ or tissue.

Chronic Dose — A dose received over a period of time, usually many days or years — as opposed to an acute dose.

Collective Dose — The collective dose is a term that has been used when the individual personal doses are added in a large population. This summed dose has sometimes been misused to estimate numbers of cancers expected in a large population, under the "Linear No Threshold (LNT)" model of radiation risks. This model is based on the assumption that any dose down to zero has a proportional risk to those estimated from the acute doses of Japanese survivors who were in a high-enough range of exposures to prompt radiation at Hiroshima and Nagasaki to be determinable. The LNT model, when translated for use to U. S. population characteristics and longer term exposures, has incorporated for chronic exposures a "Dose and Dose-Rate Reduction Factor (DDREF) of only the constant 2 (reducing acute model risks only to 1/2 for official application in radiation protection of workers or the public. As documented in Chapter II, these assumptions are not scientifically valid. The Health Physics Society scientists in the field of radiation protection have published a position paper rejecting the use of collective dose for estimating cancers or other effects in populations,

Cosmic Rays — Radiation originating from events occurring in the sun or other events in astronomical bodies that send a variety of "cosmic ray" particles into the Earth's atmosphere. As they travel down through the Earth's atmosphere these particles interact with atoms of oxygen, nitrogen, hydrogen, and other substances to produce a variety of mesons and other secondary particles. Much of the background dose to populations at ground level is due to (cosmic ray) neutrons from these reactions that cascade to ground.

Curie (Ci) — The Ci is a unit used to express amounts of radioactivity — either measured or calculated. One Ci is that amount of material containing 37,000,000,000 (37 billion) atoms transforming per second (tps or dps, or becquerels (Bq)). This number is expressed in scientific notation as 3×10^{10} Bq (or tps, or dps). Smaller subunits of the Ci are used for very large or very small quantities, by using the prefixes given in Chapter II (e.g., 1 μCi = one-millionth of a Ci; 1 kCi = 1,000 Ci).

Decay — This common word is sometimes used by radiation scientists to connote "radioactive decay"; see the definition of radioactive decay and explanations in Chapter II.

Dose — When used in discussions of radiation science or radiation protection, the word dose refers to the amount of energy deposited per unit mass at a point in tissue, or averaged over the whole body, depending on the context. Some units of dose are erg/gram, rad, rem, gray (Gy), or seivert (Sv), as explained in Chapter II.

Dosimeter — A general term applied to devices designed to record personal exposure when worn somewhere on the body (or in a humanoid phantom in testing a cancer treatment or experiment). The devices may be active or passive. Passive dosimeter is the term used for dosimeters that require processing or reading with instruments in a laboratory (such as a film packet, taken from a badge worn on the body, that needs chemical development of the film and then reading of optical densities at various points to interpret dose; or a thermoluminescence (TLD) dosimeter that needs heating in a device that also measures the glow curve of light emitted from different positions). Active dosimeters are electronic dosimeters that incorporate not only the radiation-sensing material, but also a meter circuit so that the user can at any time see the dose rate and the accumulated dose during the wearing period.

Dosimetry — Dosimetry is the general word used to refer to the science, practice, or instruments used to measure radiation dose.

Dirty Bomb — A bomb of ordinary explosive material, not a nuclear bomb, in which the explosive material has been mixed with one or more radioactive species, in order to panic a public that has not been informed — as in this book — that it is unlikely members of the public will be exposed to harmful amounts of radiation from such bombs, which are also called RDDs (radiological dispersal devices, radioactivity dispersing devices, etc. — see reference in the book that explain why few persons would likely be exposed to harmful levels of radiation from such devices used by terrorists).

Effective Dose — Effective dose (ED) is the current word used when an average absorbed dose to a limited part of the body, or just certain organs or tissues, is multiplied by factors related to organ or tissue sensitivity, so that the expressed ED will give an estimate of about the same cancer (or total) risk as if the entire body had received a dose of the given amount ED.

Exposure — An amount of radiation energy, or quantity of photons or particles, flowing toward the body per unit area, as opposed to the dose deposited within body tissue. Much of the science of dosimetry deals with data and methods of calculating doses within the body from various types of exposure to various types and energies of nuclear radiations.

Fallout — Fallout in this book means radioactive material released to the atmosphere that then falls onto the ground or onto persons or objects on the ground.

Gamma Rays — Gamma rays are very-high-frequency electromagnetic rays (bunched like particles of high energy called "photons") that are emitted from radioactive nuclei that are disintegrating (transforming), or are created when positrons (e+ particles emitted from certain nuclides) combine with ordinary negative electrons. Gammas emitted from the large variety of radioactive nuclides have wide ranges of energies. Positron annihilation, when a positron joins a negative electron, emits precisely two photons of gamma rays, in opposite directions, each photon having the energy equivalent of the mass of the electron, 0.511 MeV, which is the energy equivalent of the mass of an electron according to $E = mc^2$. The phenomena of pair production also produces positrons, when very high energy gamma rays (energy greater than 1.022 MeV) interact with an atom in matter or tissue and produce an electron-positron pair. Particles produced by this type of interaction, which increases in frequency as energy increases above 1.022 MeV and Z of the atom increases, are two 0.511 MeV photons, emitted in exactly opposite directions after a positron is slowed down in matter and annihilated. This type of radiation is the basis for the PET (positron emission tomography) scanners now used effectively in nuclear medicine to examine nutrient or drug distributions in the body that can not be seen by other scanning or radiographic equipment. In PET scanners, the part of the body being examined is surrounded by a cylinder containing many photomultipliers fed into a complex computing system. Because the two 0.511 photons emitted upon annihilation of the positron go in exactly opposite directions, the system of photocells on all sides of the body can, with the complex mathematics built into the computer system, provide an image of where the two photons originated for each event. The system can then produce an image of the distribution in body organs of where the particular diagnostic agent administered has been incorporated (e.g., a tumor?).

Gamma-Ray Constant — This is a term not needed for preparing the protective actions as described in this book. As indicated following the glossary title, it is included only for readers who might refer to information in the 2011 book. The gamma-ray constant is a value for a specific gamma-ray emitting nuclide, which is used to estimate radiation intensity at different distances, before correction for absorption by intervening air or other absorbers. These constants have been accurately calculated and confirmed by measurements of their radiation exposure intensity at a given distance from small ("point") sources. Gamma ray constants and their use for radioactive nuclides most likely available to terrorists are included in a table in Brodsky (2011), with suggestions on their use by the technically-inclined or scientists coming in for rescue or recovery.

Geiger Counter (or Geiger-Mueller (GM)) Counter — An instrument for detecting radiation and estimating radiation exposures. It uses a closed chamber filled with air or some gas at low pressure and very thin wires at high voltages (e.g., above about 1,000 volts) to detect with great sensitivity any ionizing particle entering the chamber by inducing the particles to cause an avalanche of electrons, and thus a current pulse, for each ionizing particle entering the chamber. See Chapter IV for a further description.

Genetic Effects — Effects present in the offspring or future progeny of those exposed to radiation (or any other mutagenic agents).

Gray (Gy) — A unit of dose of ionizing radiation, named after a physicist of the early 20th century who published data on densities of energy deposition by ionizing radiation that led to current theories of radiation dosimetry. 1 Gy = 1 joule of energy deposited per kilogram of tissue or other material. See discussion in Chapter II for further understanding.

Half-Life (Radiological) — The time it takes for a radioactive atom to break down until only half is left.

Half-Life (Biological) — See **Biological Half-Life.**

Health Physics — The name given to the groups of scientists, working on development of the atomic bomb in the Manhattan Project in the early 1940s, who were assigned to develop ways of monitoring radiation exposure and limiting risks to the early atomic bomb workers. It has continued to be used to connote the profession of persons dedicated to promoting safety in the use of nuclear or radioactive materials.

Health Physics Society — A professional society of persons dedicated to the science or practice of protection of workers, the public, and the environment, from harmful amounts of radiation exposure. A long history of experience has shown that the health and economic benefits of the many health and industrial uses of radioactive materials and nuclear energy may be advanced without harm to workers, the public, or the environment. The public needs to know and understand this. Members of the Health Physics Society (only about 5,000) know this and can refute exaggerated statements of radiation risks that deny us benefits to health and economic survival, even though many of them are not familiar with, or do not subscribe to, the phenomena of hormesis presented in this book and its references. Questions of radiation safety should be directed to the Health Physics Society at the website www.hps.org and to the new group Scientists for Accuracy on Radiation Information (SARI) at http://radiationeffects.org.

Hormesis — The phenomena in which an agent or chemical that causes ill health or death at high doses becomes supportive to health at doses below some level. The root of this word is the same as that of the word hormone, because the root *horm* is latin for growth related to health. See Chapter II.

Hydrogen bomb — A hydrogen bomb is also an atomic (nuclear) bomb, but one in which the two **lightest atoms, hydrogen-1 (H) and deuterium (hydrogen-2, a proton with neutron added)**, fuse together (**fusion**) as they bump into each other at the extremely high speeds caused by a surrounding fission bomb detonation (uranium or plutonium) that heats the **H-D mixture to temperatures like those on the sun.**

Explanation: The energies released by hydrogen bombs are usually thousands of times greater than those of what we are calling atomic bombs. When we talk of hydrogen bombs in this book, we are talking of the **H-D fusion bombs**. In hydrogen (fusion) bombs, energy $E = mc^2$ is released because the mass of the fused H-3 (tritium) atom **is less than the total of the H and D atoms that fuse together**. In the fission bombs (uranium or plutonium) energy is released because **the two fragments (fission products) together and the emitted neutrons weigh less than the original uranium or plutonium atom before it breaks up (fissions) in two.**

Relative explosion energy released: When talking of the impacts of what we call her atomic or hydrogen bombs, we talk in terms of the impact of comparable amounts of TNT (trinitor toluene), which everyone is familiar with. Comparisons are either in the kiloton (kT) or megaton (MT) energy release of TNT. Atomic bombs that use uranium or plutonium, like the ones at Hiroshima or Nagasaki, in about the **15 kiloton (kT), release energies** equal to about 15,000 tons of TNT, which for ton of 2,000 pounds, amounts to more than 15,000 x 2,000 = **30,000,000 pounds (30 million pounds) of TNT**. A **1 kT bomb**, which could easily be brought into a harbor and cause devastation and contamination to a large city, would be **equivalent to a 2 million pound bomb of TNT**. A **15 MT hydrogen bomb**, about like the ones I was assigned to in the Army as a physicist, releases the equivalent of **30,000,000,000 (30 billion) pounds of TNT**. See concise but complete-enough chapters by the great Drs. John Auxier (2004) and Joe Alvarez (2004).

IND — Short for "Improvised Nuclear Device," something about 1 kiloton (kT) or slightly higher in explosive energy that can be assembled by anyone using information on the internet, once the uranium is available in the right amount and the right enrichment in U-235. It is the name sometimes given to a "small" nuclear bomb. 1 kT = 1,000 tons x 2,000 pounds TNT per ton = 2,000,000 pounds (2 million pounds) of TNT — a far greater release of explosive energy than in any other type of chemical bomb, such as TNT.

Ingestion — Eating substances or taking material into the body through the mouth and gastrointestinal system.

Inhalation — Breathing in material and taking it into the lungs.

Injection — Taking material into the body through a cut or open wound.

Internal Radiation Exposure — Exposure to radiation from radionuclides deposited in tissues or organs of the body after their ingestion, inhalation, injection, or absorption through the skin Absorption through the skin has been rare, except in a few cases for water soluble substances or oils containing such atoms as tritium replacing the abundant H-1 in molecules of water, or other radionuclides imbedded in soluble substances..

Ionizing Radiation — Radiation with enough energy per quantum so that an individual quantum reaction (or hit) with an atomic electron has enough impact to eject a tightly-bound, inner-shell electron from an atom and produce an ion pair (a free speeding electron and a remaining ion of the initial atom). It is distinct from non-ionizing radiations. Non-ionizing radiations only dislodge outer, more weakly bound electrons from atoms or molecules. Examples of non-ionizing radiation health effects at high exposure levels are: blindness from direct laser light incident into the eyes; sunburn from ultraviolet rays; and blindness or neurologic effects from high intensity microwave antennae near-field radiation. The Health Physics Society has committees of specialists that recommend standards of protection against non-ionizing radiation. Further study of the nature, sources, effects and protective measures for ionizing radiation may be obtained from the authoritative texts by Cember and Johnson (2009) and Turner (2007; the text by Cember and Johnson also has details on non-ionizing radiation.

Neutrons — Neutrons are uncharged subunits of all atoms except hydrogen-1 (H-1 or 1H); the latter hydrogen atom, so ubiquitous in nature, consists of only one proton in the nucleus. For the purposes of this book in protecting the general public, a detailed discussion of neutron radiations is not given. See the definition of **Hydrogen Bomb** above for the importance of neutrons in propagating a fission avalanche in uranium or plutonium. The texts by Turner (2007) and Cember and Johnson (2009) are good sources for persons interested in detailed information about radiation sciences and health physics.

Non-Ionizing Radiation — See contrasting information in the definition of **Ionizing Radiation**.

Personal Decontamination — Personal decontamination involves the removal of radioactive materials from the clothes, skin, and hair. For purposes of this book, ordinary washing with soap and water would suffice.. It is not likely that more drastic methods would be needed, nor would they necessarily be safer without supervision by a physician.

Personal Radiation Exposure (or Dose) — Personal exposure or dose in the context of radiation protection means the exposure or dose of an individual, as opposed to the collective exposure or dose of a population of individuals. See the definition of **Collective Dose** to see the precautions of the Health Physics Society against uses of collective dose and its quantitative calculation.

Quantum — As a result of discoveries in the first half of the 19th century, both electromagnetic radiation such as gamma rays and x rays, as well as particles such as alpha and beta, are known to act in dual capacities as quanta or waves, depending on the interactions with atoms and materials under consideration. These quanta (plural of quantum) turn out to have discrete energies when emitted from nuclear reactions; this is the reason they are called "quanta." X-rays produced by ordinary x-ray machines or "atom smashers," where free electrons are smashed upon targets to produce high energy photons in ranges suitable for cancer therapy or industrial uses, do come in a continuous range of quantum energies, often called photon energies for electromagnetic radiation. In-depth understanding of these phenomena is not needed for the most important protective measures, as given in this book. The texts cited in the references are also available for the reader wanting in-depth knowledge of this science.

Rad — This is the earlier SI unit for **Absorbed Dose** of **ionizing Radiation** — the concentration of energy absorption in the vicinity of a point or in a volume of interest. 1 rad = 100 erg/gram, and also = 0.01 Gy in the more recently devised SI units. See Chapter II for the necessary understanding of absorbed doses.

Radiation — Radiation is a general term for energy in transit from a source to an absorber. In the field of health physics, the types of radiation defined under **Ionizing Radiation** or **Non-Ionizing Radiation** are the kinds of radiation considered for protecting health and the environment.

Radioactive Contamination — Radioactive contamination is radioactive material spread over some area of land, structure, material, or person. It might or might not be at a level hazardous to health or the environment. Appropriate limits or action levels need to be understood for either emergencies or lifetime exposure, in order for rational actions and decisions to avoid harm to health or the environment.

Radioactivity — The emission of alpha, beta, or gamma rays when unstable atoms break up. Units are in section II: curie (Ci), Becquerel (Bq); see summaries here in bold type.

RDD (e.g., Radioactivity Dispersing Device) — A bomb laced with radioactive material to scare people. See best information in Alvarez (2004). Unfortunately, in the public media these have been called "dirty bombs," but they are not full of germs.

Rem (roentgen equivalent man, mouse, or mammal) — The rem is the traditional SI unit of equivalent dose, as related to long-term effects of radiation. A dose to an organ or tissue in part of the body is multiplied by two factors — a radiation weighting factor and a tissue weighting factor — in order to express a resulting equivalent dose in rem that would give the same long-term chance of cancer later in life for a person of average age, as if the person had received the same dose in rad to his or her entire body. More details on this subject are not necessary for this book, but may be pursued in the texts cited for those interested in the science.

Roentgen (R) — The roentgen, named after the discoverer of x rays, is the traditional SI unit of measure for radiation exposure to x and gamma radiation in the energy ranges most confronted in radiation protection for radioactive materials in medical or industrial laboratories. A simple colloquial definition is: One R is the amount of exposure of x or gamma radiation that creates 1 electrostatic unit of charge of each sign per cc in free air of standard temperature and pressure. An exposure measurement in R has easy and safely calculated relationships to the radiation dose in air in the vicinity of a point of interest, usually the position where a part of a person's body might be when assessing worker or environmental exposures. That is, the R unit has a simple approximate relationship to the absorbed dose in a small volume of tissue at the point, and other advantages beyond the scope of this book in relating exposures to doses in various parts of the body, for fields of various shapes and sizes of incident radiation.

As mentioned in the text, official agencies have ceased using the R, and the Health Physics Society has banned its use as well as all other traditional SI units. To this author, the banning of these traditional units is a serious censoring mistake made without careful scientific and practical considerations. Such banning is a "politically correct" decision made to comply with practices in other nations that were also invoked with an apparent lack of understanding of field monitoring practices for radioactive materials with gamma emission photon energies less than 3 MeV, as opposed to the routine methods of dosimetry in vivo or in humanoid phantoms in radiology and radiation oncology where higher photon energies are now used in therapy. This ban will be a great detriment to the use in dosimetry of much data and methodology of value that was produced by earlier members of the International Commission on Radiation Units. It is particular detrimental to the education of young scientists entering the field of health physics. Dr. Lauriston Taylor, the founder of the field of radiation protection in the U.S., and an early President of the Health Physics Society, was the scientist most active in developing the concepts and measurements of the roentgen with the standard air chambers he developed at the National Bureau of Standards in the early half of the 1900s. Having known him and his efforts, I believe he would today be opposed to the banning of the traditional units. So, with the support of his memory, and many other health physicists who responded to my petition, and for the reasons in Chapter II, I intend to continue using the roentgen and related exposure-dose concepts in my books and in my dose calculations. No apologies. Amen.

Risk — As used in this handbook and as I prefer, the word risk is used to mean the same as the probability of chance that an event might occur; it does NOT imply in this handbook that the event is a harmful health effect or danger. The words, adjectives or modifiers, that precede the word risk in context should signify the event being discussed. The word "probability" is sometimes thought too deep a mathematical term, but for practical purposes means the same as "chance". The risk, or chance, of a harmful health effect or death can often be zero at low enough levels of radiation exposure; the risk might also be negative, or hormetic, within certain ranges of radiation exposure, as also explained in Chapter II. If more elementary concepts of probability and statistical phenomena were taught in elementary school using games of change that children play, graduate courses in statistics would be much easier to teach, and more adults would not fear some of these words. They apply to many events in normal living.

Shielding — Shielding by putting radiation absorbing material between sources and areas occupied by workers or the public is a major protection principle for reducing exposures to ionizing radiation. See section IVB and Exhibit 7.

Sievert (Sv) — The sievert (Sv) is the unit of biological dose currently recommended by the International Commission on Radiological Units and the International Commission on Radiological Protection to be in the Systeme Internationale (SI) system of dose and radioactivity units. Its relationship to the earlier ("traditional") SI unit of biological dose is 1 Sv = 100 rem. See further discussion in Chapter II.

Specific Activity — The specific activity of any substance is the number of atoms in the substance that are disintegrating (transforming) per second per unit mass of the entire substance. Often, in some contexts, the specific activity of a pure radioactive nuclide is cited per unit mass of the nuclide itself that is in some mixture with other substances. This latter specific activity of a nuclide is a fundamental quantity of the radionuclide itself, and is related to its half-life by a simple equation.

Stochastic Effect — An effect that occurs by chance and has a probability of occurring over a range of doses, in discussions of radiation effects. In discussions of the chances of late effects of radiation in official government reports or regulations, or in the recommendations of international expert groups, the effects, such as cancer, will be denoted as "stochastic effects." These official or international groups often give the impression that effects of short-term, acute exposures, are not chance effects, but are deterministic effects. Not so. Even acute effects occur over a range of doses, depending on the chances at any particular dose of interactions of radiations with the sensitive sites in cells, and even the chance state of health, rest, and susceptibility of the person being exposed at the particular time(s) of exposure. See Exhibit 3.

Thermoluminescence Dosimeter (TLD) — Thermoluminescence dosimeters (TLD) are among a number of devices for measuring radiation exposures in applied dosimetry. They involve the use of crystals (e.g, calcium sulfate) that fluoresce in dark chambers when exposed to heating, their fluorescence being proportional the amount of radiation dose they have received. The amount of fluorescence at each part of the TLD wafer is measured over time by a photomultiplier tube and circuit. The fluorescent glow over time is called a "glow curve." TLDs are incorporated in holders called "badges", which have multiple filters to discriminate the fluorescence in various parts of the fluorescent packet in order to determine the penetrating power of the radiation to which the wearer is exposed. Such filter system arrangements are also used in film badges.

X Rays — X rays are electromagnetic radiations emitted when collisions of atoms with impinging particles bring one or more electrons in the atom to higher shells and then electrons drop back into former shells emitting x ray photons of specific energies, or when speeding electrons in x-ray machines or atom smashers are drastically stopped by impinging on targets, or changed in energy and direction, then emitting x-ray photons over a continuous range of energies. See **Gamma Rays** for a comparison.

APPENDIX B

Methods for Checking Food and Water Concentrations for Emergency Use

This appendix is provided for the use of scientists, or members of the public with backgrounds in science teaching or other interests, who might be available to help that small portion of the public who might need to rely on food or water that might have been exposed to radioactive fallout (such as the Japanese farmer mentioned earlier who refused to evacuate from Fukushima and needed to drink milk and eat food from his farm (Englund 2011; Higgins 2011). As recommended earlier, all citizens should be able to find canned, bottled, or other foods protected from fallout within their home cabinets or refrigerators, and have sufficient supplies on hand for at least two weeks, and preferably longer as quantities for preserving life.

The following table in **Exhibit 15** of concentrations of radioactive material in food or water for consumption during emergencies includes recommended limits for 10-day consumption, based upon the fact that the food and water if exposed in an open garden or farm to fallout will NOT continue to be ingested for many months or years. Many official recommendations of government bodies (see Brodsky 2011) provide limits of contamination in food and water after emergencies, assuming they might be used for an unlimited amount of time, even perhaps a lifetime, and the general population will include infants and children who might be more susceptible than adults. Some recommendations are given after the table of concentrations below, for further reduction of these concentrations for infants or children.

EXHIBIT 15

Limits on Food and Water Concentrations for Emergencies*

The following table of 10-day emergency limits of radioactive materials is taken from Brodsky and Stangler (2004). For **simple homemade ways in which the average citizen can test food and water** against this table, with high margins of safety, **see the next page.**

Nuclide Mix	Activity (µCi cm^{-3})	Activity (Bq g^{-1})	Effective Dose[†] After 10 Days (Sv)
Gross fission product beta activity (< 30 days decay after burst) (FCDA 1955)	0.09	3,000	0.043 (using Ce-141 as surrogate)
Gross alpha activity from nuclear burst (any period after burst) (FCDA 1955) (Not likely to be detected as important in early fallout material.)	0.005	180	0.9 (Pu-239 as surrogate, but delivered over 50 years)
Phosphorus-32	0.015	500	0.024
Cobalt-60	0.03	1,000	0.068
Strontium-Ytrium-90	0.0003	10	0.0056
Mo-99-Tc-99m	0.08	2500	0.03
Iodine-125	0.03	1,000	0.13
Iodine-131	0.002	60	0.026
Cesium-137-Barium-137m	0.012	400	0.104
Iridium-192	0.03	1,000	0.028
Polonium-210	0.0006	20	0.48
Radium-226 plus daughters	0.000012	0.4	0.0022
Plutonium-238/239	0.003	100	0.50
Americium-241	0.003	100	0.40
Curium-244	0.003	100	0.24
Uranium (natural or depleted)	0.03	1,000	0.9

*Emergency levels of radioactivity in water and food for 10-day consumption.

*The values for mixed fission products and gross alpha activity after a burst are taken from a 1955 recommendation of the Federal Civil Defense Administration (FCDA), referenced in Brodsky (2011). These recommendations were derived by a group of professors expert in toxicology, and although concentration recommendations for lifetime exposure are different and have been somewhat changed over the years, the recommendations for mixed fission products are still deemed quite safe today. The values for the single nuclides deemed in a 2003 American Nuclear Society (ANS) report and by Alvarez (Brodsky 2004), plus two additional nuclides that are in widespread use in nuclear medicine, have been obtained by taking the ratios of the earlier maximum permissible concentration (MPC) for water in peacetime (Brodsky 1996) of the nuclides divided by the MPC for Ce-141 of 0.003 µCi cm^{-3} as multipliers of the values for gross beta activity in FCDA (1955) (See Brodsky and Stangler (2004)for the reasons these derivations are deemed safe). These ratios are approximately the same as obtained from the tables of Eckerman in Brodsky *et al.*(2004).

†The effective dose is the sum of the products of committed dose equivalent over 50 years multiplied by the tissue weighting factors of ICRP 60 (International Commission on Radiological Protection), as calculated by Eckerman (2004) for single nuclides, except for I-125, where the main dose is to thyroid and the committed dose from NCRP Report No. 70 (published by the National Council on Radiation Protection and Measurements in 1982) is multiplied by the weighting factor 0.05 as used for thyroid in ICRP Report 61 (published by the International Commission on Radiological Protection in 1991).

FURTHER NOTES ON EXHIBIT 15 FOR ANY SCIENTISTS USING IT IN RESPONSE OR RECOVERY PERIODS

The suggested limits for 10-day ingestion under emergency conditions may be extended for longer time periods by reducing the intakes in inverse proportion to time. Concentration values can be converted to adult intakes, if needed, by assuming a water intake of about 2,000 cubic centimeters per day, or a food intake of at least 2,000 kilocalories (Calories) per day at an energy value of about 4 Calories per gram (for combined carbohydrate, fat, and protein). Assuming a rounded density of 1 gram per cc for food results in a food intake on the order of 500 cc per day.

Note: The Japanese government has set a safe allowance level of radioactivity in foods at 100 Bq kg^{-1} (Edquist 2014). This is 1,000 times less than the emergency level above for plutonium-239, and even 4 times less than that for radium-226, even though these more radiotoxic nuclides are not among those emitted from the Japanese nuclear power plants. Although the Japanese government is undoubtedly assuming that the radionuclides to be ingested will remain in food and water for more than 10 days, the radionuclides of importance from the Fukushima reactor emissions will not have such radiological and environmental half-lives to be around in substantial quantities after several weeks.

A SIMPLE AND SAFE WAY OF TESTING FOOD AND WATER FOR 10-DAY EMERGENCY USE IN THE HOME:

For a layman with a **comparison standard** made as described below, and a simple open-window Geiger counter (which allows beta radiation to enter the Geiger tube), may be used, for any of the likely individual or combined mixtures, to check for allowable for 10-day ingestion: **An open-window Geiger counter at 1 cm from an ointment tin-sized sample (or in any inverted metal lid of, say, a 16 ounce bottle of spaghetti sauce) containing a few ounces of food or water will indicate it is safe for emergency consumption for 10 days if the Geiger counter reads less above the food or water than above the comparison standard.** (Suggested limits for alpha emitters like plutonium are also given in the above table and the book, but this information is deemed much less important for this current citizen summary.)

PREPARATION OF A COMPARISON STANDARD CONTAINING A SAFE LOW AMOUNT OF URANIUM

The beta radiation comparison standard is prepared in the lid of a 4-ounce ointment tin (7.9 cm diameter by 2.3 cm height as described in the 1955 FCDA pamphlet, but now the inverted metal lid from a 16-ounce bottle of spaghetti sauce would do) as a base to hold the sample. Each standard is prepared by mixing 3.14 grams of finely divided (60 mesh) uranyl acetate (obtained in a one-pound bottle from a chemical company) in 5 grams of liquid casting plastic, or asphalt. Other uranium oxides or salts could be used (Brodsky 2011). Thoroughly mix the liquid plastic with the uranium powder, add 6 to 8 drops of the catalyst that is purchased with the liquid plastic (in a hobby shop), stir with a glass rod for 5 minutes. Clean the ointment tin or inverted lid with detergent, rinse and dry. Then, weigh about 9 grams of the uranium-plastic mixture into the lid or tin; 9 grams would be about one-third of an ounce weighed on a postal scale. Next, heat the lid at low heat on the flat surface of a stove until the material in the lid begins to harden, then remove it from the stove. Overheating will cause the plastic to separate from the lid.

Notes on using the comparison standard with a Geiger counter to check food and water, or just using the counter even without the standard:

- First, the reader should know that the emergency recommendations in this handbook are my own, and have not been blessed by any agency of government, or any expert scientific group (although they are consistent with limits on emergency intake and levels of internal radiation dose in the 1950 recommendations of the Federal Civil Defense Administration) . More recent "Generic Action Levels" for radioactive materials in foodstuffs for public consumption after emergency release, as recommended by the International Atomic Energy Agency (IAEA) and the National Council on Radiation Protection and Measurements (NCRP), are included in Exhibit 32 of my book, and will be seen to be much lower than the 10-day (or 30-day if divided by 3) limits for the levels given. However, the IAEA and NCRP table values are for long-term intake; they will be seen to be consistent with those in the above table if the ratio of times for use of these foodstuffs were applied to the values in the above table.

- Only the first line in the table of Exhibit 15 is applicable for mixed fission products in fallout that was released from nuclear bombs (sometimes euphemistically called "improvise nuclear devices" (INDs) in media or government reports), or short-burst criticality accidents such as that at Chernobyl. These first-line values are applicable for bombs or short bursts only if the measurements are made within 30 days after detonation. After that, longer-term nuclides such as strontium-90 and cobalt-60 might become more important parts of the mixture biologically; their much lower limits of intake can be seen in the above table.

(The values in the first line are NOT applicable to single or a few specific nuclides packed into ordinary bombs (radioactivity dispersal devices (RNDs). The remaining lines in the table are applicable to the specific radionuclides indicated,)

I would also note that, if the mixture of fission products from the Three Mile Island (TMI) accident meltdown were entirely released to the atmosphere, which it was not, then the limits of intake in the first line of the above table would have been applicable only if the measurements of food and water were made in the first 3 days after release. Only less than one-millionth of the volatile iodine nuclides were released to the environment (Brodsky 1982). Exhibit 3 in Brodsky (2011) and surrounding comments should be examined by readers interested in the extreme misinformation and exaggeration of risks from TMI presented to the public by one of my former colleagues. Measurements showed that nobody in the public was exposed to harmful levels of radiation exposure from TMI. This is one of the main reasons I have needed to write my books.

➢ Lines below the first in the table are included so that individual nuclides that might be included in dirty bombs can be assessed for emergency use in food or water. The activity concentration of 0.09 µCi cm^{-3} is seen in Exhibit 33 of Brodsky (2011) to be equivalent to 180,000 disintegrations per minute (dpm) per cubic centimeter of food and water. (Note: the value of 180,000 contains a round-off error. A microcurie is 37,000 disintegrations per second (dps or Bq); multiplying by 60 seconds per minute gives the easily remembered 1 microcurie (µCi) = 2.22 million dpm. Next, multiplying by the rounded 0.09 in the table gives 199,800 or about 200,000. The rounded calculation in Brodsky 2011 gave 180,000; however, great accuracy is not of much consequence in these calculations to give approximate detection levels.

The beta emission from these disintegrations may be assumed to be the same number, and half would be emitted upward and half downward into the inverted lid. Therefore, about 100,000 betas from the food or water in the tin would be emitted upward per cc of food or water. Comparing the 0.0003 limit for Sr-90 with the 0.09 limit, the upward flow of betas from Sr-90 should be limited to about 1/300th, or 100,000/300 = about 330 betas per minute per cc of food in the lid.

The volume of an ointment tin or lid of about 8 cm diameter and 2.3 cm depth would contain the following volume: from high school geometry, $\pi r^2 d$ = 3.14 x 4 x 4 x 2.3 = 116 cc. Thus, a filled lid might emit, aside from self-absorption of betas by the food, 116 x 330 = about 38,000 betas upward per minute. Even with self-absorption, it is thus evident that a serious contamination of food or water would be detectable, even without comparison with a uranium standard. Some tests should made of these estimates by a DHS program designed for citizen self-monitoring of food and water in the absence of laboratory availability in the immediate hours and days after a disaster.

➢ Considering: the above information; the fact that any strontium-90 obtained by terrorists and further subjected to the oxidation of a bomb explosion is not likely to be taken up through the gastrointestinal lining into the circulation and deposited in the body; and the fact that 9 ounces, which would be almost 300 cubic centimeters of food or water, it is evident that an upward flux of flux of 300 x 900 = 270,000 betas per minute would not be harmful, even for strontium-90 — the single beta emitter giving the highest effective internal radiation dose to internal organs, i.e., of bone and bone marrow.

➢ Therefore, I was wrong in my calculations in Brodsky (2011); I had not considered that almost 300 grams, or cc, of food or water was being tested. I might have been too restrictive in suggesting a very conservative count rate of 1,000 counts per minute. This measurement should be tested for any radionuclides that might be used in RDDs.

➢ Caution should be used by members of the public to obtain Geiger counters or other instruments with windows thin enough to allow entry of any of the beta emissions from the nuclides listed in the above table. Manufacturers (listed in Appendix F of Brodsky (2011)) should be consulted about specifications and price before selecting an appropriate instrument by a group in a neighborhood. An adequate instrument should be available for about $500, contributed by several neighbors. If needed, training and assistance should be obtained from a high school physics or chemistry teacher in the neighborhood; many have attended courses conducted by members of the Health Physics Society over many decades in the past.

> The beta count rate of about 16,000 per minute from the uranium-coated ashtray in Exhibit 11, which is similar to the orange Fiestaware dinner plates on your grandma's table, may be compared with an estimate of 27,000 if it is assumed that only 10% of the upward betas of high energy from Sr-90-Y-90 are detected. Although these are rough estimates of count rates from different types of samples, the comparison might provide the perspective that what seems like a lot of radioactivity might still not be high enough to be a serious hazard. Radiation doses and risks must be carefully analyzed. There is no evidence that anyone was harmed by these dinner plates, used only for fractions of days over a lifetime.

*Comments: I made a uranium standard in my kitchen in the late 1950s. As bad as I am in the kitchen, if I could make this comparison standard, anyone can. It would still be safe for anyone to do; the discussion of safety is included in Brodsky (2011) along with scientific literature references up to recent times. However, I doubt that many readers will bother making this standard, except perhaps some high school chemistry teachers. The procedure for making this standard is abstracted here from my book, because there might be those who want to avail themselves of food taken from their outside garden, and they might not want to wait for, or trust, scientists to come to check their food in time. **People can die of starvation or malnutrition if food or water is improperly denied to them by overly restrictive standards that do not take into account the special emergency situations.**

I would hope that some official from DHS might read this book and recommend manufacture and provide wide distribution for such comparison standards, and the affordable kinds of Geiger counters that could easily be used to check such samples. This would help provide ordinary citizens with an introduction to concepts of the possibilities of safety and survival in the event of releases of radioactive material from nuclear detonations or industrial accidents.

APPENDIX C

Reasons for Urgency in Family Preparations to Save Lives

The following facts are quoted from Dr. Gary M. Sandquist, President of The American Civil Defense Association (TACD) (Sandquist 2013):

> We live in a dangerous and frightening world. Syria has used chemical weapons (Saran* gas — the same agent used in World War I) against its own people to maintain the Assad Dictatorship. Of greater concern to U.S. interests in the Middle East are the Iran officials (sic). It is apparent that they seek to develop nuclear weapons. It requires about 250 kilograms (550 pounds) of 20% enriched uranium to produce a simple "gun-barrel" nuclear weapon similar to that used in Hiroshima in 1945. The IAEA (International Atomic Energy Agency) reported in August 2013 that Iran has 186 kilograms.

---

*According to John D. Hoyle, a long-time devotee to civil defense: Sarin (no "a") was not used in WWI but rather chlorine, mustard and some form of picric acid. It was weaponized by the Germans in WWII. Fritz Haber et al were working prior to the war on finding more effective pesticides. That is how the nerve agents came about. Organophosphates, such as the pesticide Sevin, are just not as strong. Obviously, Sarin, Tabun and Soman were much more powerful and were weaponized.

Only another 64 kilograms will provide sufficient material for a single 20-kiloton nuclear weapon. The Iranians are adding an additional 3000 high-speed centrifuges with sufficient capacity to produce this additional enriched uranium in months."

Hassan Rouhani, the new Iranian (President) handpicked by the Ayatollahs, appears to offer moderation in Iranian diplomacy with the U.S. Significantly, this same Rouhani was a key agent in the ill-fated arms-for-hostage meeting in Tehran 27 years ago and who said in a 2004 speech to the Iranian Revolution Council, "While we were talking to the Europeans in Tehran, we were installing equipment in parts of the (uranium) facilities in Isfahan."

The world is in chaos around us. I encourage you to keep a watchful eye on current events while hastening your preparations."

Also, in an article by Karen DeYoung in The Washington Post, March 5, 2014, page A7, she quotes Prime Minister Netanyahu, who in this world along with our own President should be most concerned about Iran's nuclear program, as follows:

But if you listen to their words, their soothing words, he said, "they don't square with Iran's aggressive action." Iranian long-range missiles, he said "can strike now, or very soon, the Eastern Seaboard of the United States — Washington — and very soon after that, everywhere else in the United States."

Prime Minister Netanyahu's words are consistent with what I have heard from other authoritative sources in the United States. This gives urgency to everyone in the USA to prepare protection for themselves and families, and to try to reach our leaders with the information here and in the further comments on the next page.

In a study of Israeli and Iranian nuclear facilities, and effects of an Israeli attack on the Iran facilities, it was estimated in 2009 that Iran could have nuclear bombs within about 5 years (Toukan and Cordesman (2009). It is now about six years, and we still do not have adequate access to inspect Iran's facilities.

Also, Alvarez (2004) explains that any terrorist, even one unsophisticated in the science, can put together an improvised nuclear device (IND), really a nuclear bomb in the low kT range, if having access to some stolen uranium, in a matter of weeks. Such an IND could be carried into the USA by anyone slipping through our borders.

Yet, we are so concerned about our privacy that we fight the dedicated scientists at the National Security Agency (NSA) who only want to trace calls to detect any planned attacks that could cost our lives.

Does all this worry you? It does me.

FURTHER COMMENTS REGARDING THE URGENCY OF ATTEMPTING DESTRUCTION OF NUCLEAR WEAPONS, WHILE STILL PREPARING SUPPLIES AND SHELTERS
(March 2014)

"And he shall judge among the nations, and shall rebuke many people: and they shall beat their swords into plowshares, and their spears into pruning hooks; nation shall not lift up sword against nation, neither shall they learn war any more."

—ISAIAH 2:4

President Obama, in his State of the Union address on January 29, 2013, gave what seemed to be his assurance that nothing was "off the table" to prevent the Iranian mullahs from obtaining a nuclear weapon. Yet, negotiations are projected to take six months before an agreement is reached. Can we be sure that the Iranian leaders are not in the meantime completing their first atomic bomb, making it ready for delivery to our Capitol within the near future? Secretary Hagel said that the negotiations are "…worth the risk." I do not agree, even if the risk is only 1% or very much less, if the event being risked is the complete destruction of our nation.

It is assumed that an attack on the Iranian facilities with low kiloton nuclear block blusters would start nuclear war. This is not so, because past experience with Israel's earlier destruction of the Iraq and Syrian reactors prevented nuclear wars, and I have evidence that I have presented to certain leaders that low-kT nuclear blockbusters would not yield enough radioactive cloud or fallout exposure to hurt the good Iranian people, lest be of harm to anyone in other civilian neighborhoods (hint: see pages 100-101 of Brodsky (2011), which summarizes Brodsky and Reeves (2009)). Note that a 2 kT atomic bomb is equivalent to 2,000 tons of TNT and a ton is 2,000 pounds. Thus, a 2kT atomic (nuclear) bomb is equivalent to 4,000,000 pounds of TNT. Such bombs in bunker busters can surely completely destroy any facilities underground. Many comments from congressmen and officials in the administration indicate our leaders do not understand that Iran's nuclear facilities can be destroyed without harming Iran's citizens or anyone else from radiation. The information on pages 100-101 in Brodsky 2011 summarizes an article by Brodsky and Reeves (2009), which interprets a formerly secret Soviet report to show that a bunker buster with a 1 to 5 kT atomic bomb could destroy underground nuclear facilities without harming any public beyond about 6 miles. (Remember, a 1 kT atomic bomb is equivalent to a 2,000,000 pound TNT bomb.) Take a look at this and try to get this information to your congressman and someone in the White House.

Current discussions in Congress (2014) indicate that many in Congress, as well as the majority of the public, have yet to understand the imminent threats to us all from the Middle East turmoil. To this author, nuclear weapons brought to us through our penetrable borders or on ships are the worst among the terrible threats to our survival. Today (March 10, 2014) it was pointed out on the news that two persons with fake passports were on the airplane that went down in Asia, and that one million travelers with fake passports board commercial aircraft each year. Thus, terrorists can most easily enter our nation, and bring bombs into or near major centers of population.

I have seen the devastation of nuclear bombs with yields from a few kT to 15 MT. I also worked in, and trained responders in, radioactive fallout from nuclear weapons of the sizes likely to be used by those who hate our values and ways of living. The Afterword, following this Appendix, describes how such survival of radiation effects is not only possible, but likely, for the vast majority of citizens, even from an atom bomb explosion more than a few miles away (Connor 2014). Also, this book will better ensure readiness to survive a variety of catastrophic events.

(Chemical and biological agents must be completely avoided by immediate sheltering and personal protection, if possible. Recommendations for sheltering in place at planned indoor locations will also reduce harm from these other agents. I believe that nuclear attack is the most likely the worst to occur by terrorist actions, and preparations for surviving them will also reduce harm from other agents and natural disasters.

Further information on chemical and biological agents, and threats of communicable disease epidemics originating throughout the world, may be obtained by subscribing to the news blogs of Dr. Glenn I. Reeves at his address greeves@ara.com . Dr. Reeves is a former radiation oncologist who, after retiring from medical practice, has for a number of years consulted on effects of nuclear, chemical, and biological agents, and has collected an enormous amount of information on agents of destruction and killing, and infectious diseases around the world.)

The information on biological and chemical agents in this handbook has been kept brief, because further protections against these agents than provided here do require actions by medical professionals who are funded by State and Federal agencies involved in the nation's homeland security programs.

AFTERWORD

The Good News About Nuclear Destruction

by Shane Connor, President, KI4U, Inc.

Author's comment: The material in this appendix is taken, with permission, from one of the helpful civil defense articles and pamphlets on the website of Shane Connor (Connor 2014). Shane is a businessman I knew from his ads for a number of years, but first met only a few years ago. He recognized the unfortunate decision of Congress and the President in the mid-1990s to cancel the civil defense program, as if the world had become entirely safe with the end of the "Cold War." We now know what a terribly wrong decision that was. Shane was alert and caring enough to buy as much of the civil defense equipment that he could, before it was destroyed, to make it available so those who might need it would be able to purchase it at reasonable prices. He is one of my heroes for his foresight and recognition of the need for citizen protection in this dangerous age. If any readers are disturbed that I am pushing a particular vendor's products, they may go to the long list of vendors, listed by products in my 2011 book. That list was taken from the website of the Health Physics Society, which is given in this handbook.

With tongue in cheek, Shane provides some insights about why civil defense in the USA has taken a back seat, but now requires urgent attention so that the most lives can be saved in the event of a nuclear disaster. His facts and views are consistent with mine, and will give the reader insights to why I wrote this handbook. The most recent appearance of Shane's Afterword is presented here as given on his website:.

Originally published 8/24/2006 at WorldNetDaily. **Updated & expanded below 2/11/2014**

What possible '*good news*' could there ever be about nuclear destruction coming to America, whether it is Dirty Bombs, Terrorist Nukes, or ICBM's from afar?

In a word, they are all survivable for the vast majority of American families, *IF* they know what to do beforehand and have made even the most modest of preparations.

Tragically, though, most Americans today won't give much credence to this *good news*, much less seek out such vital life-saving instruction, as they have been jaded by our culture's pervasive *myths of nuclear un-survivability*.

Most people think that if nukes go off then everybody is going to die, or it'll be so bad they'll wish they had. That's why you hear such absurd comments as; *"If it happens, I hope I'm at ground zero and go quickly."*

This defeatist attitude was born as the disarmament movement ridiculed any competing alternatives to their ban-the-bomb agenda, like Civil Defense. The activists wanted all to think there was no surviving a nuke, banning them all was your only hope. The sound Civil Defense strategies of the 50-60's have been derided as being largely ineffective, or at worst a cruel joke. With the supposed end of the Cold War in the 80's, most Americans saw neither a need to prepare, nor believed that preparation would do any good. Today, with growing prospects of nuclear terrorism, and nuclear saber rattling from rogue nations, we see emerging among the public either paralyzing fear or irrational denial. People can't even begin to envision effective preparations for ever surviving a nuclear attack. They think it totally futile, bordering on lunacy, to even try.

Ironically, these disarmament activists, regardless their noble intent, have rendered millions of Americans even more vulnerable to perishing from nukes in the future.

The biggest surprise for most Americans, from the first flash of a nuke being unleashed, is that they will still be here, though ill-equipped to survive for long, if they don't know what to do <u>beforehand</u> from that first second of the flash onward.

Most could easily survive the initial blasts because they won't be close enough to any "ground zero", and that is very *good news*. Unfortunately, though, few people will be prepared to next survive the later coming radioactive fallout which could eventually kill many times more than the blast. However, there is still more *good news* possible, as well over 90% of those potential casualties from fallout can be avoided, *IF* the public was <u>pre-trained</u> through an aggressive national Civil Defense educational program. Simple measures taken immediately after a nuclear blast, by a pre-trained public, can prevent agonizing death and injury from radiation exposure.

The National Planning Scenario #1, an originally confidential internal 2004 study by the Department of Homeland Security, demonstrated the above survival odds when they examined the effects of a terrorist nuke going off in Washington, D.C.. They discovered that a 10 kiloton nuke, about two-thirds the size of the Hiroshima bomb, detonated at ground level, would result in about 15,000 immediate deaths, and another 15,000 casualties from the blast, thermal flash and initial radiation release. As horrific as that is, the surprising revelation here is that over 99% of the residents in the DC area will have just witnessed and survived their first nuclear explosion. Clearly, the *good news* is most people will survive the initial blast.

However, that study also soberly determined that as many as another 250,000 people could soon be at risk from lethal doses of radiation from the fallout drifting downwind towards them after the blast. (Another study, released in August 2006 by the Rand Corporation, looked at a terrorist 10 kiloton nuke arriving in a cargo container and being exploded in the Port of Long Beach, California. Over 150,000 people were estimated to be at risk downwind from fallout, again many more than from the initial blast itself.)

The *good news*, that these much larger casualty numbers from radioactive fallout are largely avoidable, only applies to those pre-trained beforehand by a Civil Defense program in what to do before it arrives.

Today, lacking any meaningful Civil Defense program, millions of American families continue to be at risk and could perish needlessly for lack of essential knowledge that used to be taught at the grade school level.

The public, and especially our children, urgently need to be instructed in Civil Defense basics again. Like how most can save themselves by employing the old *'Duck & Cover'* tactic, rather than just impulsively rushing to the nearest window to see what that *'big flash'* was across town just-in-time to be shredded by the glass imploding inwards from the delayed blast wave.

"According to the 1946 book Hiroshima, in the days between the Hiroshima and Nagasaki atomic bombings in Japan, one Hiroshima policeman went to Nagasaki to teach police about ducking after the atomic flash. As a result of this timely warning, not a single Nagasaki policeman died in the initial blast. Unfortunately, the general population was not warned of the heat/blast danger following an atomic flash because of the bomb's unknown nature. Many people in Hiroshima and Nagasaki died while searching the skies for the source of the brilliant flash."

Even in the open, just lying flat, reduces by eight-fold the chances of being hit by debris from that brief, three second, tornado strength blast that, like lightning & thunder, could be delayed arriving anywhere from a fraction of a second to 20 seconds or more <u>after</u> that initial flash.

They need to also know if in the path far downwind of fallout coming, that evacuating perpendicular to that downwind drift of the fallout would be their best strategy. They must also be taught, if they can't evacuate in time, how to shelter-in-place while the radioactive fallout loses 90% of it's lethal intensity in the first seven hours and 99% of it in two days. For those requiring sheltering from fallout, the majority would only need two or three days of full-time hunkering down, not weeks on end, before safely joining the evacuation.

This *good news* is easily grasped by most people, and an effective expedient fallout shelter can be improvised at home, school or work quickly, but only *IF* the public had been trained <u>beforehand</u>, as was begun in the '50s & '60s with our national Civil Defense program.

Unfortunately, our government today is doing little to promote nuclear preparedness and Civil Defense instruction among the general public. Regrettably, most of our politicians, like the public, are still captive to the same illusions that training and preparation of the public are ineffective and futile against a nuclear threat.

The past administrations Department of Homeland Security head, Michael Chertoff, demonstrated this attitude in 2005 when he responded to the following question in USA Today;

> Q: In the last four years, the most horrific scenario - a nuclear attack - may be the least discussed. If there were to be a nuclear attack tomorrow by terrorists on an American city, how would it be handled?
> A: In the area of a nuclear bomb, it's prevention, prevention, prevention. If a nuclear bomb goes off, you are not going to be able to protect against it. There's no city strong enough infrastructure-wise to withstand such a hit. No matter how you approach it, there'd be a huge loss of life.

Mr. Chertoff failed to grasp that most of that *"huge loss of life"* could be avoided if those in the blast zone and downwind knew what to do beforehand. He only acknowledges that the infrastructure will be severely compromised — too few first-responders responding. Civil Defense pre-training of the public is clearly the only hope for those in the blast zone and later in the fallout path. Of course, the government should try and prevent it happening first, but the answer he should have given to that question is; *"preparation, preparation, preparation"* of the public via training beforehand, for when prevention by the government might fail.

The current Obama administration also fails to grasp that the single greatest force multiplier to reducing potential casualties, and greatly enhancing the effectiveness of first-responders, is a pre-trained public so that there will be far fewer casualties to later deal with. Spending millions to train and equip first-responders is good and necessary, but having millions fewer victims, by having also educated and trained the public beforehand, would be many magnitudes more effective in saving lives.

The federal government needs to launch a national mass media, business supported, and school based effort, superseding our most ambitious public awareness campaigns like for AIDs, drug abuse, drunk driving, anti-smoking, etc. The effort should percolate down to every level of our society. Let's be clear - we are talking about the potential to save, or lose needlessly, many times more lives than those saved by all these other noble efforts combined!

Instead, Homeland Security continues with a focus primarily on...

#1 — **Interdiction** — Catching nuclear materials and terrorists beforehand and...

#2 — **COG** — Continuity of Government and casualty response <u>afterward</u> for when #1 fails

While the vital key component continues to be largely ignored...

#3— **Continuity of the Public <u>while it's happening</u>** — via proven mass media Civil Defense training <u>beforehand</u> that would make the survival difference then for the vast majority of Americans affected by a nuclear event and on their own from that first initial flash & blast and through that critical first couple days of the highest radiation threat, before government response has arrived in force.

This deadly oversight will persist until those crippling *myths of nuclear un-survivability* are banished by the *good news* that a trained and prepared public can, and ultimately has to, save themselves. More training of the public <u>beforehand</u> means less body bags required <u>afterwards</u>, it's that simple.

The tragic After Action Reports (AARs), of an American city nuked today, would glaringly reveal then that the overwhelming majority of victims had perished needlessly for lack of this basic, easy to learn & employ, life-saving knowledge.

Re-launching Civil Defense training is an issue we hope & pray will come to the forefront on the political stage, with both parties vying to outdo each other proposing national Civil Defense public educational programs. We are not asking billions for provisioned public fallout shelters for all, like what already awaits many of our politicians. We are just asking for a comprehensive mass media, business, and school based re-release of the proven practical strategies of Civil Defense instruction, a modernized version of what we used to have here, and that had been embraced by the Chinese, Russians, Swiss, and Israeli's.

There is no greater, nor more legitimate, primary responsibility of any government than to protect its citizens. And, no greater condemnation awaits that government that fails to, risking millions then perishing needlessly. We all need to demand renewed public Civil Defense training and the media needs to spotlight it questioning officials and politicians, until the government corrects this easily avoidable, but fatal vulnerability.

In the meantime, though, don't wait around for the government to instruct and prepare your own family and community. Educate yourself today and begin establishing your own family nuclear survival preparations by reading the free nuke prep primer...

WHAT TO DO IF A NUCLEAR DISASTER IS IMMINENT!

Then, pass copies of it, along with this article, to friends, neighbors, relatives, fellow workers, churches and community organizations with a brief note attached saying simply: *"We hope/pray we never need this, but just-in-case, keep it handy!"* Few nowadays will find that approach alarmist and you'll be pleasantly surprised at how many are truly grateful.

Everyone should also forward copies of both to their local, state and federal elected representatives, as well as your own communities first-responders and local media, all to help spread this *good news* that's liberating American families from their paralyzing and potentially fatal *myths of nuclear un-survivability!*

Shane Connor is the CEO of www.ki4u.com. Consultants and developers of Civil Defense solutions to Government, NPOs and Individual Families.

KI4U, Inc. has been written up in ... *New York Times* (6/13/02), *Wall Street Journal* (3/14/03 & 10/5/01), *USAToday* (6/11/02 & 7/11/02), *Washington Post* (4/13/03 & 3/16/03), *Boston Globe* (8/13/05), *SF Chronicle* (6/23/02), *Newsday* (11/24/01), *WND* (1/18/05), *IEEE Spectrum* (9/01), *Bulletin of the Atomic Scientists* (5/04), *National Defense Mag* (3/04). And, our products seen on CNN, FOX, CBS, TIME mag with radio interviews on 'Coast to Coast', NPR, Glenn Beck and numerous others, also Glenn Beck's CNN TV Show - 10/12/2006! ~ KHOU TV Interview - 5/1/2012! ~ Shipping Wars - 8/28/2012 ~ KTBC TV Interview - 1/31/2013

References

Alvarez (2004): Joseph L. Alvarez, "Defining, explaining, and detecting dirty bombs," Chapter 5 in Brodsky, Goans, and Johnson, 2004, *op cit.*, pages 69-80.

Auxier (2004): John A. Auxier, "The effects of nuclear weapons," Chapter 6 in Brodsky, Goans, and Johnson, 2004, *op cit.*, pages 81-91.

Brodsky, Beard (1960: Allen Brodsky and G. Victor Beard, Compilers and Editors, *A compendium of information for use in controlling radiation emergencies*, U. S. Atomic Energy Commission Report TID 8206 (Rev), Washington, DC, September 1960.

Brodsky (1982): Allen Brodsky, "Dose and risk information on Three Mile Island," Appendix A.6.2 in Allen Brodsky, Editor, *Handbook of radiation measurement and protection, Section A, Vol. II*, CRC Press, Boca Raton, FL, 1982.

Brodsky (1996): Allen Brodsky, *Review of radiation risks and uranium toxicity, with application to decisions associated with decommissioning clean-up criteria*, RSA Publications, Hebron, CT, 1996.

Brodsky, Goans, and Johnson (2004a): Allen Brodsky, Ronald E. Goans, and Raymond H. Johnson, Jr., Editors, *Public Protection from Nuclear, Chemical, and Biological Terrorism*, Medical Physics Publishing, Madison, Wisconsin, 2004, 832 pp. (This book, prepared as a text for the 2004 Health Physics Society Summer School, contains important contributions from over 40 national experts including practical guidance and data for saving lives.)

Brodsky, Stangler (2004b): Allen Brodsky and Marlow J. Stangler, "Rules of thumb and risks of food and water contamination," in Brodsky et al., *op. cit.*, 2004, pp. 513-517.

Brodsky, Wald (2004c): Allen Brodsky and Niel Wald, "Experience with early emergency response and rules of thumb," Chapter 20 in Brodsky et al., op. cit., 2004, pp. 335-367.

Brodsky and Reeves (2009): Allen Brodsky and Glenn Reeves, "New human data versus estimates of effects of inhaling fission product mixtures," Health Physics, Vol. 96, No. 1, pages 1-4, 2009.

Brodsky (2010): Allen Brodsky, a review of a book on the history of France's nuclear power program, *Radiance of France: Nuclear Power and National Identity After World War II*, Health Physics, Vol. 99, No. 6, pp. 814-815, December 2010.

Brodsky (2011): Allen Brodsky, *Actions for Survival: Saving Lives in the Immediate Hours After Release of Radioactive or Other Toxic Agents*," MJR Publications, Baltimore, 2011, 373 pages. See the flyer on the last page of this book. E-Books are also available on Amazon.com.

Brown (2011): David Brown, "No health effects expected in U.S. from Japan's crisis," The Washington Post, April 6, 2011.

Buddemeir (2007): Brooke Buddemeir, "Self-Indicating Instant Radiation Alert Dosimeter (SIRAD) Test Results," presented at the Fifty-Second Annual Meeting of the Health Physics Society, 8-12 July 2007 (abstract in Health Physics, Vol. 93 (Supplement), July 2007.

Buddemeir and Dillon (2009): B. R. Buddemeir and M.B. Dillon, "Key Response Planning Factors for the Aftermath of Nuclear Terrorism," Lawrence Livermore National Laboratory, Report LLNL-410067, August 2009.

Buddemeier, Valentine, Millage, Brandt (2011): B.R. Buddemeier, J.E. Valentine, K.R. Millage, and L.D. Brandt, "National Capitol Region Key Response Planning Factors for the Aftermath of Nuclear Terrorism," Report LLNL TR-512111, Lawrence Livermore National Laboratory, Livermore, California, 2011. (This is a well-illustrated report providing information for the recovery of the Washington, DC, and surrounding areas after a nuclear attack. It was sponsored by the Department of Homeland Security and its Federal Emergency Management Agency, also with scientists from Sandia National Laboratory.

Cember and Johnson (2009). Herman Cember and Thomas E. Johnson, *Introduction to Health Physics*, 4th edition, McGraw-Hill Companies, Inc., New York, 2009.

Cohen (1983): Bernard L. Cohen, *Before It's Too Late: A Scientist's Case for Nuclear Energy*, Plenum Publishing Corporation, New York, 1983.

Cohen (1992): Bernard L. Cohen, *The Nuclear Energy Option: An Alternative for the 90s*, 2nd Edition, Plenum Press, New York, 1992.

Connor (2014): Shane Connor, CEO, KI4U, Inc., http://www.ki4u.com and http://www.nukalert.com. Consultants and developers of civil defense solutions to government, military, private organizations, and families. Connor's web site also provides historic literature on the effects of nuclear weapons, and a description of the Kearney meter, which can be assembled at home by any citizen and studied to better understand how collection of electric charges are used to measure gamma, beta, and alpha radiations.

Cravens (2010): Gwyneth Cravens, *Power to Save the World: The Truth About Nuclear Energy*, Alfred A. Knopf, New York, 2007; downloaded as e-book from Amazon.com, March 14, 2014; 441 pp.

Cuttler (2013a): Jerry M. Cuttler, "Commentary on Fukushima and beneficial effects of low radiation," Dose-Response Vol. 11(4): 432-443; 2013, at: http://www.ncbi.nlm.nih.gov/pmc/articles/PMC3834738/ (A most recent article by a fine scientist I know that reviews scientific evidence of hormetic effects of radiation. The article also documents how lack of radiological emergency preparations of the Japanese authorities resulted in 1600 unnecessary deaths among the 70,000 Japanese population resulting from the improper evacuation of tens of thousands of persons who were panicked by authorities about low levels of radiation that would not cause them harm. This was consistent with my predictions near the end of Brodsky (2011).

Cuttler (2013b): Jerry M. Cuttler, "Fukushima and Beneficial Health Effects of Low Radiation," presented to the Canadian Nuclear Safety Commission, Ottawa, Ontario, Canada, June 25, 2013, accessed from http://www.nuclearsafety.gc.ca/eng/pdfs/Presentations/Guest-Speakers/2013/20130625-Cuttler-CNSC-Fukushima-and-beneficial-effects-low-radiation.pdf

Cuttler (2014a): Jerry M. Cuttler, "Leukemia incidence of 96,000 Hiroshima atomic bomb survivors is compelling evidence that the LNT model is wrong," Arch. Toxicol. Vol. 88(3), 847-848; March 2014.

Cuttler (2014b): Jerry M. Cuttler, "Remedy for radiation fear — discard the politicized science," Dose-Response 12: 170-184, available at: http://www.ncbi.nlm.nih.gov/pmc/articles/PMC4036393/

Edquist (2014) Britt Edquist, "CSU Student Branch Attends Fukushima Ambassadors Program," Health Physics News, Volume XLII, Number 3, March 2014: 6-7. Hps.org accessed March 3, 2014. This issue also has many examples of the gross exaggeration of human radiation effects of the Fukushima radioactivity releases, as on the social internet sites as well as the media. The authors in this Health Physics Society are competent scientists who specialize in promoting safety and do not under-rate real dangers. There were likely no immediate or long-term deaths from the radiation releases, only from the terrible devastation of the tsunami. Those who promote untrue exaggerations of risks are either not competent in the relevant sciences, or have been have been whipped into religious fervor by decades of media misinformation. You will find them rejecting the contents and discussions of radiation effects, positive and negative, in this book.

Englund (2011): Will Englund, "At Chernobyl, warnings for Japan," The Washington Post, April 4, 2011.

Goldberg (2009): Jane G. Goldberg, *Because People Are Dying*, Sea Raven Press, Franklin, Tennessee, 2009. (Dr. Goldberg is not a physicist and misuses some of the physical terms describing radiation dose and risk, but provides an extensive review of reports by competent scientists I know on processes involved in radiation risks and hormetic effects.)

Hatfill, Orient(2013): Steven J. Hatfill, M.D. and Jane M. Orient M.D., "Immediate bystander aid in blast and ballistic trauma," Journal of American Physicians and Surgeons 2013;18:101-104. This journal issue has other articles useful for optimizing preparations to survive disasters. Information for obtaining this journal is available from www.jpands.org or (800) 635-1196.

Higgins (2011): Andrew Higgins, "Wrestling with a threat they can't see," The Washington Post, April 5, 2011

Hiserodt (2005): Ed Hiserodt, U*nderexposed: What If Radiation is Actually GOOD for you?* Laissez Faire Books, Little Rock, Arkansas, 2005. (This little book is written tongue-in-cheek with some sarcastic humor to make important points, but, although there are a few misuses of physics terms, the book is scientifically correct at the level required for the non-scientific members of the population, and has been developed by this engineer Hiserodt after extensive literature review and many interviews with outstanding professionals, whom I know to be among the best nuclear scientists and engineers.)

ICRP (1991). International Commission on Radiological Protection, *ICRP Publication 60, 1990 Recommendations on Radiological Protection*, Annals of the ICRP 21(1-3), 1991.

Jones (2014): Steve Jones, training cards, available from Doctors for Disaster Preparedness, c/o Dr. Jane Orient, 1601 N. Tucson Blvd., Suite 9, Tucson, AZ 85716-3450. **Kirk Paradise, Emergency Plans Coordinator for Huntsville, Alabama, and Pat Bersie of the Utah Division of Emergency Management wrote the original versions of this card.**

Luckey (1991): T. Donald Luckey, *Radiation Hormesis, 2nd Edition*, CRC Press, Inc., Boca Raton, FL, 1991.

Rockwell (2003): Theodore Rockwell, *Creating the New World and Images from the Dawn of the Atomic Age*, with a Foreword by Dr. Glenn T. Seaborg, Nobel Prize-Winning chemist, 1st Books Library, Bloomington, Indiana, 2003.

Raabe (2011)): Otto G. Raabe, "Toward improved radiation protection standards," Health Physics Vol. 101(1):84-93; 2011.

Sandquist (2013): Gary M. Sandquist, "President's Message," Journal of Civil Defense, Vol. 46: 2; 2013. Information on the availability of this journal is at www.taca.org . Issues from 1968 to present have many articles for in-depth preparations by survival-minded persons willing to devote the necessary time and effort.

Taylor (1980): Lauriston S. Taylor, "Some nonscientific influences on radiation protection standards and practice," Health Physics Vol. 39: 851-874; 1980.

Toukan and Cordesman (2009): Abdullah Toukan and Anthony H. Cordesman, *Study on a Possible Israeli Strike on Iran's Nuclear Development Facilities*,

Center for Strategic and International Studies, Washington, DC, March 14, 2009.

Turner (2007): James E. Turner, *Atoms, Radiation, and Radiation Protection*, 3rd edition, WILEY-VCH Verlag GmbH and Co., 2007.

University of Pittsburgh (2011): Monica Shoch-Spann, Ann Norwood, Tara Kirk Sell, and Ryan Morhard, *RAD RESILIENT CITY: A Preparedness Checklist for Cities to Diminish Lives Lost from Radiation After a Nuclear Terrorist Attack*, Center for Biosecurity of the University of Pittsburgh Center for Health Security, August 2011, accessed March 10, 2014 at www.radresilientcity.org.

Walchuk and Wahl (2014): Mary Walchuk and Linnea Wahl, "Fried by Fukushima: Misunderstanding, Misinformation, and Misapprehension; What HPS Can Do," Health Physics News, Vol. XLII, Number 3, March 2014.

Waltar (2004): Alan E. Waltar, *Radiation and Modern Life: Fulfilling Marie Curie's Dream*, Prometheus Books, Amherst, New York, 2004.

Index

A

A Bomb, 3, 97, 100
Absorbed Dose, 8, 9, 96, 107, 109
Absorption, 9, 10, 107
Acute Dose, 19, 21, 96, 99
Acute Effects, 14-18, 111
Alpha Particles, 96
Alvarez, Joseph L., 64, 105, 108, 115, 123, 134
American Nuclear Society, 73, 86, 115
Anastas, George, 72
Anderson, Rip, 85
Association of American Physicians, 31, 88
Atomic Age, 138
Atomic Energy Commission-National Research Council, 145
Atomic Energy Commission, 74, 134, 145
Auxier, John A., 64, 105, 134
avalanche, 51, 52, 97

B

Background Radiation, 26-29, 53, 97
BACnet, 59
Beard, G. Victor, 134
Beck, Glenn, 133
Becquerel, 11, 97, 108
Bersie, Pat, 138
Beta Particles / Beta Rays, 7, 48, 52, 55-56, 97
Biological Dosimetry, 98
Biological Half-Life, 12, 98, 103
Biological Terrorism, 134
Biosecurity, 139
Blow-Out Kit, 63
Boston Globe, 133
Boston Marathon, 61
Bq, 11-13, 97, 99, 100, 108, 115, 118
Brandt, L. D., 135
Brown, David, 92, 135
Buddemeir, Brooke R., 15, 26, 38, 135
Bush, G. W., 40, 41

C

Canadian Nuclear Safety Commission, 136
Cancer, 6-24
Cember, Herman, 21, 106, 136
Cesium-137, 12, 28, 97
Chernobyl, 14, 17, 27, 42-44, 79, 117, 137
Chertoff, Michael, 130, 131
Chronic Dose, 19-25, 99
Civil Defense, 1, 30, 31, 34, 37, 41, 47, 51-53, 60, 71, 72, 76, 77, 86, 115, 117, 121, 127-133, 136, 138, 143
Cohen, Bernard L., 82, 84-86, 90, 136

Cold War, 36, 72, 127, 128
Collective Dose, 99, 107
Common Radiation Doses, v, 26, 27
Connor, Shane, iv, 4, 47, 58, 95, 125, 127, 133, 136
Cordesman, Anthony H., 123, 139
Cosmic Rays, 97, 99
Cramer, Beth, 55
Cravens, Gwyneth, 70, 81, 84-86, 88, 90, 94, 136
Curie, 11, 13, 99, 108, 139
Cuttler, Jerry M., 20, 23, 24, 29, 41, 42, 44, 45, 50, 70, 80, 88, 93, 134, 136, 137

D

Department of Defense, 41
Department of Homeland Security, 15, 30, 40, 59, 129, 130, 136
Department of Radiation Health, 74
DeYoung, Karen, 122
Dillon, M. B., 38, 135
Dirty Bomb, 42, 43, 46, 100
Dirty Bombs, 1, 35, 39, 64, 108, 118, 127, 134
Disaster Preparedness, 138
Dose, 96-98, 99
Dose-Rate Reduction Factors, 99
Dosimeter, 15, 39, 40, 42, 44, 46, 48, 49, 56, 57, 100, 111, 135
Dosimeters, iii, vi, 4, 15, 16, 33, 39-42, 45-47, 55, 56, 66, 80, 82, 95, 100, 111
Dosimetry, 9, 75, 98, 100, 101, 103, 109, 111

E

Edquist, Britt, 70, 87, 115, 137
Enewetok Island, 68
Enewetok, 20, 66, 68

F

Fallout, 101, 113, 117, 124, 125, 128-132
Family Emergency Plan, 30-39, 121
Federal Civil Defense Administration, 47, 115, 117
Federal Emergency Management Agency, 47

FEMA, 47
fission, 12, 29, 35, 64, 77, 78, 80, 90, 97, 104, 106, 115-117, 135, (see atomic bomb, 97)
Fukushima Daichi, 70, 86, 87
Fukushima, 29, 41-43, 50, 59, 70, 71, 80, 81, 86, 87, 92, 95, 113, 115, 136, 137, 139
fusion, 104 (see hydrogen bomb, 104)

G

Gamma Rays, 11, 28, 52, 101-102, 107, 108, 112
Gamma, v, 7, 8, 10-12, 19, 21, 28, 35, 39, 41, 42, 48, 50-56, 58, 67, 68, 98, 101, 102, 107-109
GBq, 11-13
geiger counter, 33, 49-53, 55, 58, 59, 64, 67, 79, 103
Geiger-Mueller Counter, , 49-53, 103
Generic Action Levels, 117
Genetic Effects, 103
GM Counter, (see Geiger-Mueller Counter)
Goans, Ronald E., 73, 134
Goldberg, Jane G., 20, 86, 137
Gotchy, Reginald, 78, 92
Gray, 9, 12, 96, 100, 103
Gus'Cova, A. K., 17
Gusev, I., 17
Gy, 9, 10, 13, 14, 17, 18, 21, 96, 100, 103, 107

H

H-1, 106
H-3, 104
H-bomb (hydrogen bomb), 56, 57, 104
Haber, Fritz, 121
Half-life, 11, 12, 103, 111
Hansen, Rick, 45
Hatfill, Steven J., 31, 61-63, 65, 137
Health Physics Society, 103-104
Hg (symbol for mercury), 63
Higgins, Andrew, 29, 80, 113, 138
Hiroshima-Nagasaki, 77

Index

Hiroshima, 19, 36, 39, 54, 70, 87, 97, 99, 105, 121, 129, 130, 137
Hiserodt, Ed, 20, 86, 138
Homeland Security Committee, 1, 2, 72
Homeland Security, 1, 2, 5, 15, 30, 40, 59, 72, 82, 95, 125, 129, 130, 132, 136
Hormesis, 18-24, 104, 138
Hormetic, 19-24, 44, 67, 86, 87, 93-95, 104, 110, 136, 137, 138
Hoyle, John D., 121
Hydrogen Bomb, 20, 28, 36, 65, 66, 68, 104-105
Hypothermia, 63

I

IEEE Spectrum, 133
Inhalation, 3, 12, 92, 98, 105
Injection, 98, 105
Interdiction, 132
Internal Radiation Exposure, 105
International Atomic Energy Agency, 117, 121
Iodine-131, 12, 29
Ionization Chamber Survey Meter, 53, 54, 60
Ionizing Radiation, 11, 42, 51, 54, 103, 106-108, 110
Iranian Revolution Council, 122

J

Johnson, Raymond H., 69, 73, 74, 87, 89, 134
Johnson, Thomas E., 136
Jones, Steve, 1-3, 45, 138
Journal of American Physicians, 31, 137
Journal of Civil Defense, 30, 138
JP Laboratories, 4, 46

K

Kaku, Michio, 82
Key Response Planning Factors, 135
KI4U, 34, 38, 49, 50, 58, 60, 127, 133, 136
kiloton, 39, 105, 124, 129
Knealle, 75

Knopf, Alfred A., 136
kT, (see Kiloton)

L

Late Effects of Radiation, 19-25
Lawrence Livermore National Laboratory, 135
Long-term Effects of Radiation, 19-25
Luckey, T. Donald, 20, 138

M

Makeshift Shelter, 34
Mancuso-Stewart, 74-78
Mancuso, Thomas, 74-76, 78, 80
Manhattan Project, 103
MBq, 13
McQueary, Charles E., 15, 40, 41
megaton, 36, 105
Mettler, Fred., 17
Millage, K. R., 135
Miller, Ken, 79
millirem, 6-11, 18
Minogue, Robert, 83
Morhard, Ryan, 139
mR (a unit of exposure), 6-13, 108-109
mrad (a unit of dose), 6-13, 54, 67, 100
mrem (a unit of dose), 6-13, 27, 28, 100
mSv (a unit of dose), 6-13, 27, 43, 44, 46, 100
MT (megaton of TNT), 105, 124

N

Nagasaki, 19, 36, 39, 70, 87, 97, 99, 105, 130
National Bureau of Standards, 109
National Security Agency, 123
neutron, 19, 54, 68, 104, 106
NOAA Weather Radio, 31
Non-Ionizing Radiation, 106, 108
Norwood, Ann, 139
Nuclear Bomb, (see atomic and hydrogen bomb)
Nuclear Energy, 70, 74, 85, 86, 89, 90, 94, 103, 136

Nuclear Power, 3, 12, 14, 27, 43, 46, 69, 70, 72, 77-83, 85-88, 90, 94, 115, 135
Nuclear Regulatory Commission, 79, 83, 84, 91
Nukalert, 58-59, 136

O

Office of Nuclear Regulatory Research, 83
Ohno, Kazumoto, 80
Orient, Jane M., 31, 61, 62, 65, 137, 138

P

Paradise, Kirk, 138
Patel, Gordhan, 4, 40, 42, 44-46, 49, 82
PERDS, 45
Personal Decontamination, 32, 66-68, 107
Personal Radiation Exposure, 34-48, 55-57
Personal Radiation Monitor, 39-48
Planned Shielding Materials, 34-39
plutonium, 8, 24, 29, 104-106, 114-116
pocket chamber, 47-48
pocket ionization chamber, v, 33, 47, 48
Potassium-40 (K-40), 28
preparedness checklist, 30-33, 139
Prime Minister Netanyahu, 122

R

R (unit for roentgen), 13, 108-109
Raabe, Otto G., 24, 138
rad, 8-13, 107-108
Radiation Dose, 8-13, 107-108`
Radiation Effects Research Foundation, 87
Radiation Health Department, 28
Radiation Health, 28, 71, 74, 76
Radiation Hormesis, 19-24, 104, 138
Radiation Information, 6, 88, 104
Radiation Protection, 103, 107-109 (see also Health Physics)
Radioactive Contamination, 26, 39, 79, 108
Radioactivity Dispersing Device, 108
Radioactivity, 11-13, 108

RADSticker, 44
RADTriage-FIT, 46
RADTriage, 44, 45, 55-57
Ranges of Radiation, 26-29, 55-59
RDD, 3, 43, 67, 92, 108
Reeves, Glenn I., 124, 125, 135
rem, 9-10, 13-18, 108
RERF, 87
Risks of radiation, 6-25, 110-111
RNDs, 117
Rockwell, Theodore, 20, 84-86, 90, 138
roentgen, 6-8, 108-110
Rouhani, Hassan, 122

S

Sanders, Barkev S., 74-76
Sandquist, Gary M., 121, 138
Saran, 121
SARI (Scientists for Accurate Radiation Information), 86, 88, 104
Sarin, 121
Seaborg, Glenn T., 138
Selenium, 20
Self-indicating Instant Radiation Alert Dosimeter, 39-46, 135
Sell, Tara Kirk, 139
Shoch-Spann, Monica, 139
SI Prefixes, 13
sievert, 8, 10, 110
Sinclair, Warren, 19
SIRAD Dosimeters, 4, 15, 33, 39-46, 55, 56, 82
SIRAD, 4, 15, 33, 39-42, 48-46, 49, 55-57, 82, 135
Smith, Phillip, 50, 58
Stangler, Marlow J., 40, 114, 115, 135
Sternglass, Ernest J., 13, 73, 76-80, 89, 91
Stewart, Alice, 75, 76, 80
stochastic effects, 19-25, 111
Sv, 10, 18, 19, 21, 27, 70, 87, 100, 110

T

Table of Approximate Densities and Thicknesses for One-Fifth Reduction, 35
Taylor, Lauriston S., 19, 23, 88, 89, 109, 138
Tehran, 122
Terrorist Nukes, 127
The American Civil Defense Association, 31, 121
The China Syndrome, 70
Thermoluminescence Dosimeter, 111
Thermoluminescence, 100, 111
Three Mile Island, 13, 14, 27, 77-79, 91, 117, 134
Three-Mile-Island, 12
Toukan, Abdullah, 139
Trainer, Jennifer, 82
Turner, James E., 106, 139

U

U. S. Atomic Energy Commission, 134
U. S. Nuclear Regulatory Commission (NRC), 83
Ukraine, 42, 43
units of radiation, 6-13
uranium, 56, 104-106, 114, 116, 119-123, 134

Utah Division of Emergency Management, 138

V

Valentine, J. E., 135
Van de Graaff, 14

W

Wahl, Linnea, 70, 73, 86, 139
Waite, David, 83
Walchuk, Mary, 70, 73, 86, 139
Wald, George, 83
Wald, Niel, 14, 15, 17, 28, 29, 76, 135
Waltar, Alan E., 69, 70, 84-86, 90, 139
World War I, 121
WWII, 121

X

X Rays, 107, 108, 111

Z

Zero Ranges of Radiation Dose and Risk, 5

ABOUT THE AUTHOR

Dr. Allen Brodsky was educated with a B.E. in chemical engineering from The Johns Hopkiins University (JHU, 1949), an Atomic Energy Commission-National Research Council (AEC-NRC) Fellowship in Radiological Physics at the Oak Ridge National Laboratory (1949-50), a masters in physics (JHU, 1960), and a doctor in science (Sc.D.) in biostatistics and radiation health, University of Pittsburgh (UPitt,1966). He is certified by the  American Board of Health Physics (CHP, 1960), the American Board of Industrial Hygiene (CIH, Radiological Aspects, 1966), and the American Board of Radiology (Dipl. 1975, Radiation Therapy Physics).

In addition to experience establishing radiation safety programs at the Naval Research Laboratory and several universities, writing regulations and radiation safety guides for the Atomic Energy Commission and the Nuclear Regulatory Commission, developing radiation treatment plans for cancer patients at Mercy Hospital in Pittsburgh, and conducting radiation measurements and radioepidemiologic studies, he has taught radiation sciences and mentored graduate students at UPitt, Duquesne University School of Pharmacy in Pittsburgh and Georgetown University and Catholic University in Washington, DC. Of special pertinence to this book are: his experiences in the Army as a physicist measuring prompt radiation signals from the first three hydrogen bomb detonations at Enewetok in 1952-54; as a physicist in the Federal Civil Defense Administration (FCDA) establishing training programs for responders and civil defense officials in 1956-57, and training responders in atomic bomb fallout fields in Nevada in 1957; negotiating for the AEC the first AEC-DOD joint operations center for responding to radiation accidents; as a Research Associate, Associate Professor (UPitt) and Technical Director of Radiation Medicine (Presbyterian-University Hospital, 1964-70), designing the whole body counting and bioassay laboratory and evaluating patients who had accidentally inhaled plutonium, americium, uranium, and various fission products; and as the first Ad Hoc Chair of Homeland Security for the Health Physics Society (HPS) after 9/11 in 2001-2002.

In addition to journal articles and documents written for research projects and government regulatory agencies, this handbook is his ninth book published for commercial distribution — written to provide information to the public to save lives in radiation emergencies and other types of life-threatening natural and man-made disasters. His other books are described on www.BrodskyBooks.com.

Among his awards for teaching, research, and safety practice are: the Robley Evans Medal, Founders Award, and Fellow Award of the HPS; the Radiation Science and Technology Award of the American Nuclear Society, the Distinguished Graduate Award of the Graduate School of Public Health of UPitt, and the Vicennial Award of Georgetown University.

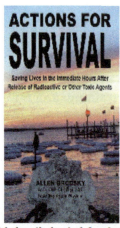

NECESSARY READING FOR RESPONDERS AND CITIZENS

Actions for Survival

Saving Lives in the Immediate Hours After Release of Radioactive or Other Toxic Agents

Allen Brodsky

373 pages, 50 Exhibits, Notes, Bibliography
$44.95 Suggested Retail Price *(See sale price below)*

This book provides the rationale for, and necessary life-saving information for, preparing the general member of the public for personal and family survival. It also gives recommendations for "top leaders" of government agencies and emergency response organizations to enhance their abilities to protect the public or reduce effects from weapons of mass destruction or effluents from nuclear or chemical accidents. The emphasis in this book is to provide the simplest information and rules of thumb for early actions in the seconds, minutes, and early hours after an attack, before the limited number of scientists are able to assess extents of damage and exposure, and before the limited number of responders will be able to assist individual families and members of the public. The book also explains how the public has been subjected to exaggerations of low-level exposures, provides information on radiation risks versus dose, and recommends public education that would prevent unwarranted panic in the vast areas that might be contaminated at lower levels that do not pose significant risks. Early projections of risks of radiation from the Japanese Fukushima nuclear reactors are included. The book is divided into the following chapter and appendix headings:

Chapter 1 -   An Unpleasant Truth
Chapter 2 -   Why Prepare Yourself for Personal and Family Protection
Chapter 3 -   A Few New Words to Remember
Chapter 4 -   Some Basic Information About Radiation Risks
Chapter 5 -   Basic Information about WMD That Can Save Life
Chapter 6 -   Actions for Those Who Do Not Want to Bother with Much Preparation
Chapter 7 -   Additional Preparations Before an Attack That Would Reduce Risks
Chapter 8 -   The Importance of Wearing Self-Reading Pocket Exposure Monitors (or "Pocket Dosimeters")
Chapter 9 -   Summary of Life Saving Actions for Laymen and Leaders
Appendix A – More Word Definitions and Concepts for Further Reading
Appendix B – A Reprint of Presentations Given In Early Civil Defense Training
Appendix C – Poster: Actions That You, The Public, Can Take for Protection in a Terrorist Attack
Appendix D – A Sample of Ted Rockwell's Socratic Answers to Radiation Myths
Appendix E – Summary of Problems with the New SI Units in Regard to USA Homeland Security
Appendix F – List of Companies Selling Instruments and Protective Equipment Which Can Be Helpful in Personal Preparations
Appendix G– Copies of Purchasing and Specification Information for Self-Indicating Colorimetric Personal Monitors
Appendix H – Department of Homeland Security Planning Guidance
Appendix I –   Late Breaking News
Afterwords
References
About the Author

Allen Brodsky, Sc.D., CHP, CIH, DABR, educated with a B.S. in chemical engineering, an AEC-NRC Fellowship in Radiological Physics at Oak Ridge National Laboratory, a master's in physics, and a doctorate in biostatistics and radiation health, is certified by the American Boards of Health Physics, Industrial Hygiene, and Radiology (Therapeutic Physics). In addition to experience establishing radiation safety programs at the Naval Research Laboratory and several universities, writing regulations and radiation safety guides for two Federal agencies, developing radiation treatments for cancer patients, and conducting radiation measurements and radioepidemiologic studies, he has taught radiation sciences, and biostatistics and epidemiology, and been a mentor to over 150 graduate students at three universities. Of special pertinence to this book is his unique combination of experiences as a scientist measuring prompt radiations at the first three hydrogen bomb tests in the Pacific; as a physicist at the Federal Civil Defense Administration establishing training programs for responders and civil defense authorities; as trainer of responders in fallout exercises at nuclear tests in the Nevada Test Site; as a negotiator for the Atomic Energy Commission (AEC) in 1957 of the first joint AEC-DOD operations center for responding to radiation accidents; as a professor and Technical Director of Radiation Medicine managing and evaluating patients exposed to external radiation and internal deposition of plutonium, americium, and fission products in the early growth industries in Pittsburgh in the 1960's; and as the first Chair of the Ad Hoc Committee on Homeland Security of the Health Physics Society in 2001-2002. In addition to his journal articles and many documents published for the government, this is his eighth book, providing data and methods for measuring or estimating internal and external exposures, and avoiding health risks, under emergency or routine conditions. He is a founding member of the Health Physics Society and two of its chapters. Among his awards for contributions to teaching, research and radiation safety practice are the Robley Evans Medal of the Health Physics Society, the Radiation Science and Technology Award of the American Nuclear Society, the Vicennial Medal of Georgetown University, and the Distinguished Graduate Award of the Graduate School of Public Health, University of Pittsburgh.

Take advantage of our special sale: $39.95
Order From: http://www.mjrpublications.com/New-Actions-For-Survival.html
(FREE SIRAD RADIATION DOSIMETER INCLUDED)